'Anyone who is not shocked by quantum mechanics hasn't understood it'

— Niels Bohr

Contents

Preface

A hundred years ago, the world of science was upended by a theory so profound and so powerful that it has entered the popular lexicon: *quantum*. Within a few years of its formulation, quantum physics explained the nature of matter and forces across the universe, from subatomic particles to stars, resolving mysteries that had baffled scientists for decades. It has proven to be the most successful scientific theory in history, and underpins chemistry, particle physics, materials science, molecular biology and much of astronomy. It has penetrated every major industry, from mining to healthcare, and spawned several new ones. Its technological applications have shaped much of the modern world, and given us electronics, computers, AI, the internet, global communications, precision navigation, nanotechnology, LEDs, high-definition TV screens and smartphones. Lasers, transistors, superconductors and microchips are all products of the quantum age. It is no exaggeration to say that the quantum revolution, which began in earnest in 1925, was the most disruptive technological transformation in history.

But what makes quantum theory of such immediate importance over a century after its inception is that a second great quantum revolution is now underway, one that promises to rival the first in its far-reaching scientific, industrial and social ramifications. Dubbed Quantum 2.0, it is known more formally as quantum information science. It arises from the ability of scientists and engineers to control individual atoms, electrons and photons, and to process, store and transmit information in novel and previously unimagined ways. This emerging technology has led to totally unbreakable encryption protocols, sensors and scanners of unprecedented power and feats like teleportation that seem little short of magic. Above all, quantum information science holds the tantalizing promise of a completely new concept in technology – the quantum computer – that will far outperform the world's best conventional supercomputers. A fully functional quantum computer will possess unrivalled capabilities for genetic analysis, drug design, climate modelling, market analysis and the fabrication of smart materials. But the most disruptive application of a quantum computer is its ability to break the codes in widespread use for secure data encryption, a looming threat that has been dubbed the quantum apocalypse. At this time, the world is in thrall of the disruptive effects of AI. But the science that gave us AI – quantum physics – looks set to merge with it in the near future, enormously enhancing its power and scope. The impact of QAI – Quantum Artificial Intelligence – is almost impossible to predict. Just as Quantum 1.0 drove the information age that defined much of the twentieth century, so Quantum 2.0, of which QAI represents but one component, will shape the twenty-first century, likely in ways that we cannot yet imagine.

Given the promise of quantum information science, it is no surprise it is attracting the attention of governments and businesses worldwide. The UK has established a National Quantum Technologies Programme, explaining that, 'Our lives and economies will soon be transformed by quantum technologies as profoundly as they have been by steam, electric traction, radio communications and electronics.'[1] The US government passed the Quantum Initiatives Act in 2018, while President Xi Jinping has called for 'a great leap forward' in quantum technologies. Other countries have followed suit.[2] UNESCO has declared 2025 to be the International Year of Quantum Science and Technology. Companies too are scrambling to exploit second-generation quantum technology in the financial, mining, healthcare, energy and aerospace industries, with applications across most sectors of the economy. In their third annual Quantum Technology Monitor, the investment firm McKinsey assess that quantum technology could create market value worth trillions of dollars within the next decade.[3]

With opportunities inevitably come threats. In 1994, it was discovered that a quantum computer could be programmed to crack many of the world's standard encryption codes used to protect financial transactions and confidential information exchange. What emerged was a new arms race, especially between China and the West, to develop quantum computers first. Big tech firms soon piled in and are currently leapfrogging each other to attain 'quantum supremacy'. Quantum sensor technology is also advancing in leaps and bounds. It is making conventional military detection and navigation systems obsolete, and raising the stakes for smart surveillance of submarines, missiles and stealth bombers. With quantum sensing, today's lurking menace could be tomorrow's sitting duck. Combining quantum target detection with AI threatens to completely disrupt the global strategic landscape. And, while quantum sensors are proving a boon to medical science, projects that use quantum neuro-imaging to couple brains to computers have triggered unsettling ethical concerns. As with all disruptive technologies in their infancy, the downstream consequences are hard to predict, but it is safe to say that whoever controls Quantum 2.0 controls the world.

It may then come as a shock to learn that, for all its momentous impact on our lives and well-being, quantum physics stems from a theory that, to put it bluntly, makes no sense. That theory – called *quantum mechanics* – works brilliantly, but it implies that the atoms, molecules, electrons and photons that are so profitably manipulated by scientists and engineers do not actually have a definite independent existence. The concrete world of daily experience dissolves away at the atomic level into a blurry amalgam of blended realities, resolved only by the intervention of some form of specific measurement. Common sense and intuition fail completely when we try to grasp what is 'really going on' in the quantum domain. On its inception in the 1920s, quantum mechanics proved so troubling that some of the world's leading physicists, most notably Einstein, flatly rejected its implications of fuzzy reality and laboured to discredit the theory in its accepted form. To do this, they zeroed in on a prediction of the theory so weird – so 'spooky', in Einstein's words – that it seemed to be self-evidently absurd: namely, that particles on opposite sides of the lab, or even the galaxy, are entangled by a sort of telepathic tether in a way that has no counterpart in normal life. It took until the 1980s for an experiment to be performed to test the prediction, and to the consternation of the sceptics, it demonstrated that nature is just as spooky as Einstein feared.

What followed was remarkable. It dawned on physicists that this long-range 'entanglement' could be embraced and used as a resource to drive a new technological era – what matured into today's quantum information science. Rather than sweeping the weirdness under the carpet, as was done for decades, scientists and engineers took it at face value and began leveraging it for practical applications.

The successes of this approach notwithstanding, merely accepting the intrinsic ambiguity of quantum reality hasn't solved the fundamental problem, which is this: how does the focused reality of the everyday material world emerge from the blurred amalgam of its microscopic constituents? The quest to understand what, precisely, it means to *know* something about a quantum particle remains an outstanding challenge. There are many contending interpretations of quantum mechanics, some of them invoking parallel universes or mental as well as physical states. These philosophical deliberations have been accompanied by a flurry of additional experimental tests suggesting that the quantum world is even weirder than we thought.

My aim in writing this book is to explain in basic terms what quantum mechanics is and how it works, tracing it back to its origins in early twentieth-century physics. I will outline its stunning technological implications and carefully discuss what it says – and doesn't say – about the deep nature of reality. In this endeavour I am hampered by the number one rule of science popularization: no mathematics allowed! Many of the weird aspects of the quantum world cannot be properly described in everyday language, only in equations and symbols. I have done my best using analogies and informal descriptions, but that inevitably risks creating misunderstandings. Since quantum mechanics remains a work in progress, with sharp disagreements among professionals about interpretational aspects, the reader should be cautioned that I have put my own slant on the subject. I have included a Bibliography for those who wish to explore alternative approaches and ideas.

PART ONE

Paradigm Shift

CHAPTER 1

The Birth of the Quantum Concept

Suppose that the everyday world we experience through our senses is but a tiny, impoverished fragment of a stupendously greater realm, one that is incomparably rich, exuberantly dynamic and bewilderingly alien. Envisage a domain of boundless possibilities, of subtle convolutions of form and substance, distributed all around us and inside us, flowing out into infinite dimensions beyond our ken and even beyond the reach of our imagination. That dazzling, gargantuan domain is in fact the world in which we are already inescapably embedded, but from which we are almost totally shut out. Access to this alien world is attained only through infinitesimal portals, observational pinholes that afford but momentary glimpses of a seething wonderland of restless activity, vaster than all the universe we see, vaster even than all possible universes we can comprehend, vaster indeed than all conceivable vastness.

Welcome to the quantum universe.

And immediately we hit a foundational question: is this magical metaverse merely an abstraction, a mathematical frolic of interest only to physicists and philosophers, or does it in some sense *really exist*? That question – what is real? – goes right to the heart of the quantum story, and indeed of the entire scientific enterprise.

To ease ourselves into this weighty topic, let's start with a brief anecdote. When I was fifteen, my sister's boyfriend saw a ghost. At least, that's what he claimed, having spent the night sleeping downstairs in our living room, which we children long believed was haunted. My mother was sceptical, however. 'He was probably just dreaming,' was her opinion, 'or drunk.' I preferred the ghost story. Who was right? Was there 'really' a ghost in our living room or was the whole affair just a figment of the young man's imagination?

Each of us experiences a world 'out there', which we view through our eyes and interpret via the information-processing taking place in our brains. Reality, for any one of us, is a product of the external and internal, of matter and mind. To get around this, a long list of philosophers, from Aristotle and René Descartes to John Locke and Thomas Nagel, defended the notion of *objective* reality – a physical universe that exists independently of our individual observations, and which has done so since long before human observers appeared on the scene. It is a world made of material objects that move and change in response to various forces. That, at least, is the normal view of existence adopted by Western thought. And yet, as I shall explain, quantum mechanics confounds this simplistic version of an objective, independent reality. There is indeed a world 'out there', but it is far stranger than most people imagine, or, in fact, *can* imagine.

Reality is a slippery concept, about which entire volumes have been penned by eminent thinkers over the centuries. Most of us, however, make do with a rough-and-ready notion that goes something like this: if an entity is real, then its existence could, in principle, be confirmed by someone else.

It should be 'independently verifiable by a disinterested investigator', to put it formally. It's a sentiment well encapsulated in the motto of Britain's national scientific academy, the Royal Society: *Nullius in verba* – Take nobody's word for it. For the two and a half centuries after the Society's founding in 1660, that pragmatic assumption accorded with the common-sense view that there are definite facts about the world, whether or not anybody is checking. Thus, science took as its basis objective truth, as opposed to personal subjective experience that might include dreaming, hallucination, hypnotic suggestion, mirages and, well, ghosts. It therefore came as a bombshell to scientists when, early in the twentieth century, a new theory emerged, which shattered this comforting belief by implying that the external world lacks definite objective existence when it isn't being watched.

The equation that changed the world

On 27 January 1926, the world changed for ever. It was on that day that the German scientific journal *Annalen der Physik* published a paper by the Austrian physicist Erwin Schrödinger (see Fig. 1) which demolished centuries of belief about the nature of matter and the way the physical world is put together. It had been apparent for a quarter of a century that something was very wrong with the traditional concept of material objects on the atomic and molecular scale, but it took the publication of a specific equation – Schrödinger's equation – to sweep away the old picture of matter and open the path to an entirely new way of thinking about the physical universe. What emerged, following a few years of frenzied

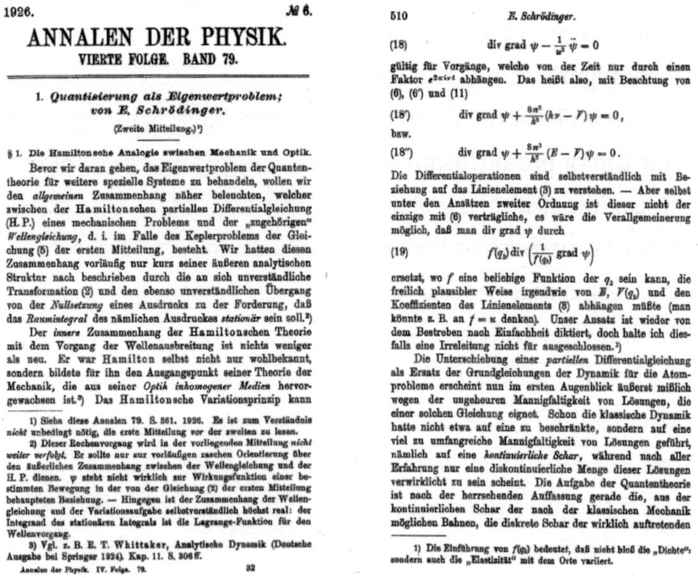

1926.   № 6.

ANNALEN DER PHYSIK.
VIERTE FOLGE. BAND 79.

1. *Quantisierung als Eigenwertproblem;*
von E. Schrödinger.

(Zweite Mitteilung.)[1]

§ 1. Die Hamiltonsche Analogie zwischen Mechanik und Optik.

Bevor wir daran gehen, das Eigenwertproblem der Quantentheorie für weitere spezielle Systeme zu behandeln, wollen wir den *allgemeinen* Zusammenhang näher beleuchten, welcher zwischen der Hamiltonschen partiellen Differentialgleichung (H. P.) eines mechanischen Problems und der „zugehörigen" *Wellengleichung*, d. i. im Falle des Keplerproblems der Gleichung (5) der ersten Mitteilung, besteht. Wir hatten diesen Zusammenhang vorläufig nur kurz seiner äußeren analytischen Struktur nach beschrieben nach die an sein unverständliche Transformation (2) und den ebenso unverständlichen Übergang von der *Nullsetzung* eines Ausdrucks zu der Forderung, daß das *Raumintegral* des nämlichen Ausdruckes *stationär* sein soll.[2]

Der *innere* Zusammenhang der Hamiltonschen Theorie mit dem Vorgang der Wellenausbreitung ist nichts weniger als neu. Er war Hamilton selbst nicht nur wohlbekannt, sondern bildete für ihn den Ausgangspunkt seiner Theorie der Mechanik, die aus seiner *Optik inhomogener Medien* hervorgewachsen ist.[3] Das Hamiltonsche Variationsprinzip kann

---

1) Siehe diese Annalen 79. S. 361. 1926. Es ist zum Verständnis *nicht* unbedingt nötig, die erste Mitteilung vor der zweiten zu lesen.

2) Dieser Rechengang wird in der vorliegenden Mitteilung *nicht weiter verfolgt*. Er sollte nur zur vorläufigen raschen Orientierung über den äußerlichen Zusammenhang zwischen der Wellengleichung und der H. P. dienen. ψ steht nicht wirklich zur Wirkungsfunktion einer bestimmten Bewegung in der von der Gleichung (2) der ersten Mitteilung behaupteten Beziehung. — Hingegen ist der Zusammenhang der Wellengleichung und der Variationsaufgabe selbstverständlich höchst real: der Integrand des stationären Integrals ist die Lagrange-Funktion für den Wellenvorgang.

3) Vgl. z. B. E. T. Whittaker, Analytische Dynamik (Deutsche Ausgabe bei Springer 1924) Kap. 11. S. 306 ff.

Annalen der Physik. IV. Folge. 79.   32

---

510   *E. Schrödinger.*

(18) $\operatorname{div} \operatorname{grad} \psi - \frac{1}{u^2} \ddot{\psi} = 0$

gültig für Vorgänge, welche von der Zeit nur durch einen Faktor $e^{2\pi i \nu t}$ abhängen. Das heißt also, mit Beachtung von (6), (6′) und (11)

(18′) $\operatorname{div} \operatorname{grad} \psi + \frac{8\pi^2}{h^2}(h\nu - V)\psi = 0$,

bzw.

(18″) $\operatorname{div} \operatorname{grad} \psi + \frac{8\pi^2}{h^2}(E - V)\psi = 0$.

Die Differentialoperationen sind selbstverständlich mit Beziehung auf das Linienelement (3) zu verstehen. — Aber selbst unter den Ansätzen zweiter Ordnung ist dieser nicht der einzige mit (6) verträgliche, es wäre die Verallgemeinerung möglich, daß man div grad ψ durch

(19) $f(q_i) \operatorname{div}\left(\frac{1}{f(q_i)} \operatorname{grad} \psi\right)$

ersetzt, wo f eine beliebige Funktion der q_i sein kann, die freilich plausibler Weise irgendwie von E, $V(q_i)$ und den Koeffizienten des Linienelements (3) abhängen müßte (man könnte z. B. an $f = u$ denken). Unser Ansatz ist wieder von dem Bestreben nach Einfachheit diktiert, doch halte ich diesfalls eine Irreleitung nicht für ausgeschlossen.[2]

Die Unterschiebung einer *partiellen* Differentialgleichung als Ersatz der Grundgleichungen der Dynamik für die Atomprobleme erscheint nun im ersten Augenblick äußerst mißlich wegen der ungeheuren Mannigfaltigkeit von Lösungen, die einer solchen Gleichung eignet. Schon die klassische Dynamik hatte nicht etwa eine zu beschränkte, sondern auf eine viel zu umfangreiche Mannigfaltigkeit von Lösungen geführt, nämlich auf eine *kontinuierliche Schar*, während nach aller Erfahrung nur eine diskontinuierliche Menge dieser Lösungen verwirklicht zu sein scheint. Die Aufgabe der Quantentheorie ist nach der herrschenden Auffassung gerade die, aus der kontinuierlichen Schar der nach der klassischen Mechanik möglichen Bahnen, die diskrete Schar der wirklich auftretenden

---

1) Die Einführung von $f(q_i)$ bedeutet, daß nicht bloß die „Dichte", sondern auch die „Elastizität" mit dem Orte variiert.

Figure 1

Two pages from Erwin Schrödinger's world-changing paper in the journal *Annalen der Physik* – the title page and the page on which his famous equation makes its first appearance (the three equations labelled 18; different but mathematically consistent forms).

analysis by the world's leading physicists, was the birth of an entirely new scientific discipline called quantum mechanics. It transformed our understanding of reality, reshaped the landscape of science, and gave birth to entire new industries that have powered economic growth for decades.

Quantum mechanics didn't spring ready-made into Schrödinger's mind. His equation was an attempt to grapple with a plethora of oddities and paradoxes that pervaded the world of pre-quantum – now known as *classical* – physics. These oddities hinted that something was seriously amiss with scientists' understanding of matter and forces: impossible stars, atoms that shouldn't exist, heat radiation that ought to incinerate the universe – a puzzling array that nobody understood until it all fell into place in the mid-1920s.

Schrödinger was not alone in his endeavours. At that time, others were grappling with the baffling internal structure of atoms – most notably Werner Heisenberg, who was taking a different mathematical approach from Schrödinger, yet one that eventually turned out to be equivalent.[1]

For readers unfamiliar with the early history, here is a brief selection of issues that baffled the world's leading scientists during the late nineteenth and early twentieth century.

Six unexplained oddities

1. ATOMS SHOULD COLLAPSE

If there is one scientific fact known to almost everyone on the planet, it is that the world is made of atoms. The basic idea goes back two-and-a-half millennia, to the Greek philosopher Democritus, who had in mind indestructible microscopic lumps that could move around and stick together to make all the material stuff in the universe. But, if atoms are so very tiny, how can we verify their existence? It's true that for centuries, the atomic theory of matter was just a conjecture. Indeed, right up to the turn of the twentieth century there were sceptics. 'I don't believe that atoms exist!' asserted Ernst Mach (he of the Mach numbers, for describing aircraft speed) as late as 1897. But the tide of history was against him. Today it's possible to image, and even manipulate, individual atoms using special equipment; their existence is no longer in doubt. But, as we shall see, the nature of that existence is very odd indeed.

With the benefit of hindsight, it's clear that a powerful clue came from the discovery of particles of electricity. The phenomenon of electricity was known from antiquity, but its nature remained a mystery until the nineteenth century. In 1854, a German glassblower named Heinrich Geissler made a precursor of the familiar fluorescent lamp – a tube containing a pair of electric terminals and a gas at very low pressure. Geissler's boss, the physicist Julius Plücker, discovered that when a high voltage was applied across the terminals, electricity would flow through the near-empty tubes in an unknown form. In 1879, the English physicist William Crookes, also using glass tubes with low-pressure gas, demonstrated that electricity emanating from the negative terminal – the cathode – caused a phosphorescent coating on the glass at the far end of the tube to glow when the invisible stream of what were then called 'cathode rays' impacted it. Crookes managed to bend the rays with a magnet, suggesting to him that it was composed of negatively charged particles. Crookes was right, and today we call the particles electrons.

But it took until 1909 for the existence of electrons to be fully accepted, when the American physicist Robert Millikan managed to trap small numbers of them on tiny oil drops and measure their individual charges.

The discovery that electricity is carried on tiny particles of matter presented an obvious problem. Given that normal matter is electrically neutral, there must be positively charged particles too, to balance things out. Before long, physicists settled on the familiar model of an atom as electrons whizzing around a compact nucleus with a positive charge. And at this point a hitch became apparent. When electric charges change speed or direction, they radiate energy in the form of electromagnetic waves. This is the principle of the transmitter, in which electrons oscillate back and forth in an antenna to generate radio waves. By the same principle, electrons going at high speed around atoms should also radiate energy, with the consequence that they would spiral into the nucleus in very short order. But when atoms do emit electromagnetic waves (e.g. from light bulbs) they don't destroy themselves in the process. Something must prevent atoms from entering a death spiral by stabilizing the electrons' orbits. But what?

2. RUNAWAY HEAT

In the 1850s, physicists deduced that there is a sort of democracy principle (known technically as the equipartition principle) at work in a hot object, whereby the total heat energy gets shared equally, on average, among all the components, that is, its constituent molecules. (This is precisely true only when the body is in thermodynamic equilibrium.)

Experiments confirmed this, but there was a puzzle when it came to radiant heat. Think of a kiln at a high temperature. Some of the heat is in the jiggling molecules of the walls and some is in the glow coming from the walls. That glow – that radiant heat – consists of electromagnetic waves: undulations in an all-pervasive electromagnetic field. The democracy principle says that each electromagnetic wave should possess about the same average energy as a molecule in the walls. But how many waves are there? Whereas molecules are discrete, countable lumps of matter, the electromagnetic field is continuous, and there seems to be no bound on the number of possible waves you could have in a kiln. In particular, the population of short wavelength (high frequency) waves grows in number without limit, because you can pack many more small waves than big waves into the same space. If all these waves got their fair share of heat energy, they would dominate everything: more and more energy would be concentrated into higher and higher frequencies – smaller and smaller wavelength waves – a problem dubbed 'the ultraviolet catastrophe'. But experiments showed that although the energy in radiant heat starts out rising with frequency, it peaks at a definite value (about 37GHz for a typical porcelain pottery kiln at 1,300°C) and then tails off rapidly at higher frequencies (= shorter wavelengths). This is illustrated in Fig. 2 (where frequency is shown increasing from right to left). If the bath of heat radiation in the kiln were like a gas made up of particles rather than continuous waves, the problem would go away.

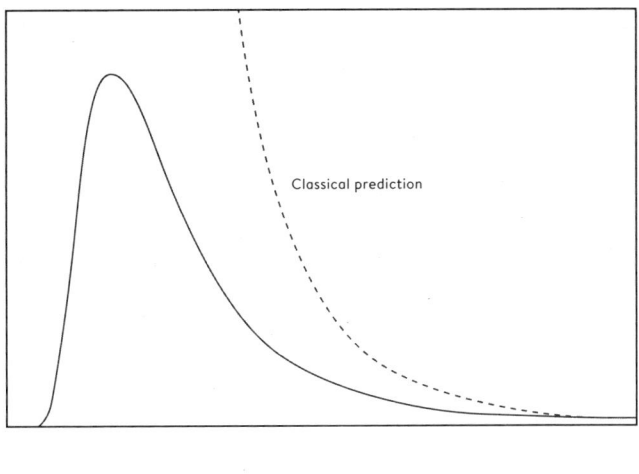

Classical prediction

Radiation intensity

← Frequency

Figure 2

The graph shows the spectrum of radiation from a heat source such as the inside of a kiln. The vertical axis is the intensity of the radiation. From right to left, the curve rises to a maximum and then falls away sharply at high frequencies. The broken line shows the prediction from classical electromagnetic theory: a curve that rises without limit at high frequencies. This obviously absurd prediction was dubbed the ultra-violet catastrophe.

3. THE PHOTOELECTRIC EFFECT

If you shine light on a metal, it makes electricity by knocking out electrons, which can then flow away freely. But the details were puzzling to nineteenth-century physicists. High frequency (blue) light has more energy than low frequency (red) light, so you might expect blue light to displace more electrons than red light. It doesn't. It ejects a similar number, but each ejected electron possesses a greater energy.

4. SUPERCONDUCTORS

Electrons flow through metals when a voltage is applied to drive them forward, but they have to batter their way through the conducting material, which is typically a sea of quivering positive ions. The obstacle race produces a resistance to the current as the electrons bounce off the ions and scatter in different directions. If the temperature of the metal is raised, the ions jiggle faster, which adds to the resistance. Conversely, when the temperature is lowered, resistance gradually falls. In the early years of the twentieth century, refrigeration technology advanced enormously. The world's foremost practitioner was the Dutch physicist Kamerlingh Onnes, who in 1908 succeeded in liquifying helium gas. He then began investigating the electrical properties of mercury (a conductor) at very low temperatures, and in 1911 made a totally unexpected discovery. When the temperature fell below $4.15\,^{\circ}\text{K}$ ($-269\,^{\circ}\text{C}$), resistance to electricity abruptly and mysteriously dropped to zero, producing a perfectly conducting state. Onnes described the phenomenon as 'extraordinary' and pronounced it to be a new 'superconductive' state. Albert Einstein described it as 'absolutely inexplicable'.

5. WHITE DWARF STARS

Curiously, at about the same time that Onnes was puzzling over superconductivity, his compatriot, the Dutch astronomer Ejnar Hertzsprung, was wondering about the colours of stars. Even a casual glance at the night sky reveals that there are red stars and blue stars and yellow stars like the sun. Hertzsprung found a simple relationship between the colour and the intrinsic luminosity of the majority of stars: they all lie near the same line on the graph of colour versus luminosity. There are dim red dwarfs at one end of the line and bright blue giants at the other. (The rule of thumb for stars is bigger = brighter = hotter = bluer.) However, astronomers noticed some peculiar stars that didn't fit anywhere near this line. These stars are very faint, suggesting they are small, but instead of being red, like red dwarfs, they're blue-white, indicating a very high temperature. The peculiar stars were dubbed white dwarfs. Although they have about the same mass as the sun, they are compressed to roughly the size of the Earth. At this sort of density, the grip of gravity is ferocious. On the surface of a white dwarf, a person would weigh 100,000 tonnes. How, then, do these compact objects shore themselves up against the crushing effect of gravity? No force known at the time could counter the star's immense gravitational pull.

6. RADIOACTIVITY

In what became a famous story in the history of science, in 1896 the French physicist Antoine Henri Becquerel accidentally discovered that the element uranium emits 'rays', after seeing that some photographic plates which he had left in a drawer next to uranium salts became 'fogged'.[2] Further experimentation showed that these fog-inducing rays were in fact helium nuclei (usually called alpha particles), ejected from the uranium nuclei. How? What was the nature of this process? It seemed completely mysterious at the time.

The first quantum leap

The first step the scientific community took towards solving these mysteries – and the day that the whole quantum saga began – came at the turn of the twentieth century. If you were to draw a line in history between 'before' and 'after' one of the most tumultuous transformations in science, this, I believe, would be it. Though it was a full quarter century before Schrödinger published his famous equation marking the true, or official, birth of quantum mechanics, looking back we can piece together the precursory steps that led up to it.

The genesis of the basic quantum concept was in 1899. It was just another frustrating day at work for Max Planck, the German physicist who took the first, decisive, leap. Planck had for years struggled to understand the problem of runaway energy at high frequencies in heat radiation – the ultraviolet catastrophe illustrated in Fig. 2. He had long been wrestling with the equations for thermodynamics and electromagnetism in an attempt to combine them in a way that matched the measured spectrum of heat radiation, which rises to a peak and then falls in a distinctive manner. In the end, the solution to the conundrum was actually disarmingly simple, banal even, and for Planck totally gratuitous. He tried out what would happen if radiant heat energy couldn't be emitted in any amount, but only in discrete little packets, which he called quanta* – deriving from the same Latin root as 'quantity', or 'portion': this is the origin of the term. Each quantum, Planck proposed, possessed an energy in proportion to the frequency of the wave: higher frequency, more energy. That required Planck to introduce a new fundamental constant of nature into physics, one that fixes the scale, giving us the actual value of a quantum of energy for a specified frequency. It is, unsurprisingly, referred to as Planck's constant. This was a drastic step, and Planck implemented it reluctantly, in the manner of a mathematical fudge – a computational manoeuvre – rather than as a serious hypothesis about nature. But it worked. If radiant heat behaves like a gas made up of particles rather than continuous waves, sharing out the energy democratically is straightforward.

Here is a rough explanation of how Planck's proposal solves the problem. Since the energy going into an electromagnetic wave is not, according to Planck, infinitely divisible, the minimum amount a given frequency could possess is one quantum. However, a wave of that frequency might possess no energy at all, i.e. no quantum. The democracy principle is just a statistical distribution, so it's okay if some wave frequencies don't have any energy, while others have one, two, three quanta of energy, and so on, so long as the average comes out right. Obviously, the higher the frequency (requiring more energy per quantum, according to Planck's

---

* A quantum of heat (and light) radiation is usually called a photon. In popular parlance, a quantum leap often means a big change, whereas most photons are extremely small units of energy.

hypothesis) the smaller the fraction of waves that, on average, can receive one quantum, even fewer two, etc. The upshot of this requirement is that the high frequency waves are suppressed in energy. It's a bit like a lottery in which just one person gets the first prize of $100, 10 people get the second prize of $10 each and 100 people get the third prize of $1. That is equipartition of sorts: each category of prize winners receives the same $100 in total. A simple calculation shows that, with Planck's proposal, the energy spectrum of heat radiation at a given temperature matches the shape of the curve shown in Fig. 2. Voilà! Problem solved, just like that – at the stroke of a pen – a bit of elementary algebra. But of course, the problem was far from solved. Planck was right, and the spectrum now bears his name in recognition, but in science it's not enough to say, 'I've been at this for years and I'm sick of the fact that the equations won't give the right answer, so I'll change one of them.' What that amounts to is altering a basic law of nature, a drastic step that you have to justify. Planck himself was less than ecstatic: he described his quantum hypothesis as 'an act of despair'. In fact, he never came to fully accept quantum mechanics, the theory he helped incubate.

The next act of the drama came when Einstein offered a neat explanation for the photoelectric effect. That was 1905, Einstein's *annus mirabilis*, in which he laid the foundations for not merely one revolution in physics, but three. Similar to Planck's suggestion that heat radiation is emitted in discrete quanta, Einstein also suggested that light comes in discrete packets, with energy proportional to their frequency. (Radiant heat and light are both forms of electromagnetic radiation, differing only in frequency.) When a stream of these light quanta – photons – hits a metal surface, electrons are kicked out, one electron per photon. If, as hypothesized by Planck and Einstein, higher frequency photons have more energy than lower frequency, then they will produce faster-moving electrons, but not more electrons. If you turn up the intensity of the light (use more photons), then you do get more electrons liberated – a stronger electric current – as observed in experiments. It was actually this work, rather than his famous theory of relativity, that later earned Einstein the Nobel Prize in Physics. Yet – as we shall see – Einstein, like Planck, never came to fully accept quantum mechanics.

The final piece of the jigsaw came in 1913. By this time, the 'quantum' idea of energy coming in discrete packets was out there, although it was by no means generally accepted, given it was still completely unexplained. The Danish physicist Niels Bohr borrowed the basic concept and proposed that the orbits of electrons in atoms are also 'quantized', that is, they are confined to certain discrete radii and energy levels as they go around the nucleus. Bohr's suggestion was not totally ad hoc. If one accepts with Planck and Einstein that light comes in little discrete packets, then the source that emits or absorbs the packets (e.g. atoms) must presumably also have discrete energy levels. In this picture, an electron may jump from one level to another, corresponding to a photon being emitted or absorbed. This early skirmish with atomic structure was a primitive model that made no attempt to account for *why* quantized energy levels exist, but Bohr proposed a formula for the energies of the levels that strikingly conformed with some known experimental results about the light spectra of hydrogen and helium. Bohr suggested that atoms would not collapse because there was a definite smallest orbit with the lowest energy – called the *ground state* – below which the electron couldn't go, and any further radiation emission was forbidden – for what at the time were unknown reasons.

Taken together, the proposals of Planck, Einstein and Bohr set the scene for what would eventually become a fully worked-out theory of quantum mechanics. But their various proposals left a lot to be desired. The fact that light is 'quantized', that it comes in discrete lumps of energy, flew in the face of overwhelming evidence that light is a wave. This caused great confusion, for how could something be both a wave and a particle? Later, physicists would refer to 'wave-particle' duality, but for two decades nobody could really wrap their heads around the idea. Nor could the existence of quantized energy levels in atoms, and in particular the existence of a lowest level, or ground state, be reconciled with any known principles.

Then came a new development. In 1924, a French aristocrat, the splendidly named Louis Victor Pierre Raymond, 7th Duc de Broglie, proposed in his PhD thesis that electrons could sometimes behave like waves. It was a bold concept, given that electrons are quite obviously little particles. According to Planck and Einstein, light waves can sometimes behave like particles, and conversely, according to de Broglie, particles of matter can also behave like waves.

De Broglie's matter-wave hypothesis was a huge leap, but he was spot on. It took a few years for his proposal to be definitively confirmed experimentally, but in the meantime, building upon de Broglie's basic idea, Schrödinger presented his famous equation to describe how the mysterious matter waves propagate (Fig. 1). He gave the strength (known as the amplitude) of the wave the Greek symbol psi, written ψ, in his original paper and the convention stuck. Schrödinger proposed that the wavelength – the distance between successive peaks of the wave – depends on the particle's momentum. For example, fast-moving electrons should have shorter wavelengths – be more bunched up – than slow-moving, less energetic, electrons. And experiments confirmed that was right. This was the decisive breakthrough, following which everything fell into place.

Electron waves readily explained the mystery of atomic energy levels. Quivering wave patterns envelop the nucleus, vibrating at distinct frequencies like the notes of musical instruments, each pattern corresponding to a specific energy level. The numbers match up brilliantly, recovering Bohr's ad hoc formula and more. The patterns have different shapes and sizes; the simplest pattern is a compact symmetric cloud that hugs the nucleus most closely; that is the ground state, below which there are no more available patterns on offer. Schrödinger's equation could also be used to describe unbound electrons colliding with atoms, scattering off them or being captured, or moving through crystals, or bouncing away from each other. In addition, it described other particles, including entire atoms and molecules, accurately predicting their vibrational and rotational motions and the forces that bind them. Applications of Schrödinger's equation came so thick and fast it was said that even a second-rate physicist could do first-rate work. The stage was set for what was to become the Golden Age of physics.

Mingled with the heady excitement, however, was a deep sense of unease, not to say outright bewilderment, concerning the sweeping implications for the nature of reality. Schrödinger's equation requires us to abandon the simplistic assumption that the atomic world is merely a scaled-down version of the everyday world and to accept that it is something else entirely, something defying common sense.

Quantum mechanics would become the dominant scientific story of the twentieth century: a sensationally successful branch of physics with applications across all the sciences and engineering, but supported by a detailed theory that seemed to strike at the very root of rational inquiry about the physical world. It was the epitome of the 'paradigm shift', according to which true revolutions do not merely change a few technical details: they reconfigure the entire conceptual foundations of the subject. The reason quantum mechanics was so hard to grasp, and remains controversial to this day, is because it demands that we part with long-cherished notions about the way the physical universe is constructed and our place within it.

CHAPTER 2

Nature's Lottery

It is perhaps no coincidence that Schrödinger published his equation at the exact mid-point of the 1920s – the 'Roaring Twenties', as the decade came to be known. Sandwiched between the horrors of the First World War and the social disruption and economic collapse of the 1930s, the twenties are often romanticized as a time of joyful experimentation and intellectual ferment. Think of modernism, surrealism, cubism and more; of painters like Pablo Picasso and Salvador Dalí, and writers such as Virginia Woolf and James Joyce, who pushed the boundaries of convention. Across Europe and America, contemporary art, architecture, music and entertainment challenged accepted norms. It was in this 'anything goes' atmosphere that scientists set about reconceptualizing the physical world at the most basic level. How, they asked, could particles such as electrons behave like waves?

The existence of matter waves was not in itself outside the scope of physical theory. After all, physicists had long enjoyed a familiarity with waves of various sorts – water, sound, radio, or seismic waves. But these are all examples of the periodic movement of a substance or energy field spread out in space, and the equations that describe how they behave involve mathematical variables which stand for tangible physical quantities – things that can be directly measured, such as the difference in height between the peaks and troughs of ripples on a pond. Electron waves, however, seemed to be something else entirely; the waves might wiggle around and spread themselves out in space, but electrons are undeniably point-like particles and not some sort of disseminated 'electron stuff' sloshing about like water. It followed that the variable ψ in Schrödinger's equation couldn't represent the value of a familiar physical quantity. It had to stand for something altogether more abstract. But what?

Waves of what?

It took another year after the appearance of Schrödinger's seminal paper for the answer to emerge. In December 1926, the German physicist Max Born* published a convincing argument suggesting that electrons are not waves of material substance as such, but of *probability*. This was a momentous departure from standard physical theory and, as we shall see, it had far-reaching consequences. The essential idea is that where the ψ wave is pronounced – where it has a large amplitude – is where the corresponding particle is most likely to be found; where it is weak, the chances are low that the particle is located there. Thus, the wave's distribution doesn't tell you where the particle is situated specifically, it only yields the betting odds for finding the particle at this-or-that location. It is a probability wave, not a physical wave. The

---

* Incidentally, Max Born was Jewish and fled the Nazis in 1933, eventually to settle in Edinburgh. His granddaughter was the famous Australian singer and actor Olivia Newton-John.

actual position of the particle is uncertain until a measurement is made; it's just a cloud of possibilities. There is a familiar analogue to probability waves in daily life. When police say that a crime wave is spreading across a city, they don't mean there is some sort of 'criminality stuff' rippling through the urban landscape; rather, there is an enhanced probability that a crime will be committed where the wave is pronounced.

Born not only provided the nexus between the wave and particle aspects of matter, he gave a simple mathematical rule for how to calculate the relative probabilities for the particle to be found at each location, given the precise form of the wave. His formula is known as the Born rule, and I shall often refer to it.* With that in hand, it became a matter of straightforward mathematics to solve Schrödinger's equation for the system of interest, and then to use Born's rule to determine the probability distribution for locating the particle in various regions of space.

Here's a simple example. First, note that the waves don't need to be wiggly and spread out: they can form localized humps, like tsunamis. Figure 3(a) shows a possible form of a hump-like wave in one dimension; it could be static, or moving. If an individual particle such as an electron is described by a wave of that shape, the most likely place for it to be found is at the middle, where the wave amplitude is a

---

* I should stress a critical technical detail. Born related the strength (or amplitude) of the matter wave at each position to the probability of locating the particle there. In point of fact, the Born rule states that you have to take the square of the amplitude to get probabilities. And to be completely accurate, one has to take the modulus squared because the amplitude of the wave is, in general, complex valued, that is, it is at each point a complex number, not a real number.

(a)   (b)

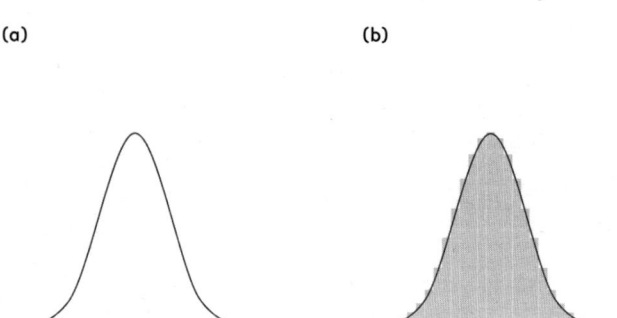

Figure 3

A possible wave form for an electron. The curve in 3(a) is a measure
of the probability for finding the electron at any particular place
along the horizontal axis. It will most likely be found near the peak
of the curve. Figure 3(b) shows typical results from many position
measurements on an ensemble of identical waves, expressed as a
histogram, which will approximate the predicted probability curve.
The example is restricted to a single dimension, but an analogous
shape can be constructed in three dimensions.

maximum. But there's a chance it could be found in one of the wings of the curve. To make this clearer, suppose someone performs a measurement to determine where the particle *actually* is. It will probably show up not far from the centre of the hump. Now suppose the process is repeated – not with the same particle, that's important, because that one has been disturbed by the measurement – but with another particle described by an identical wave. (Physicists have techniques to prepare differently shaped waves.) Next time, the position measurement will turn up a different location for the particle. The time after that, different again. If the process is repeated many times, the accumulated results track the shape of the curve as shown in the histogram Fig. 3(b). Which inevitably prompts a pressing question . . .

But where is the particle *really*?

It is at this point that we hit the issue that makes quantum mechanics so weird and controversial. Suppose the particle is found on measurement to be, say, at the location marked X in Fig. 4. An obvious question is whether the particle was there all along, lurking at X, even before the observation was made. Or, if the particle moved to X from somewhere else, then where was it, say, one microsecond before the measurement and what were its movements in the interim? According to the standard interpretation of quantum mechanics, the question 'where was the particle before its position was measured?' *has no answer*. It's not that we don't happen to know the answer or we lack the technology to track the particle's movements, it is that the question itself is deemed to be meaningless. There is simply *no fact of the matter* about

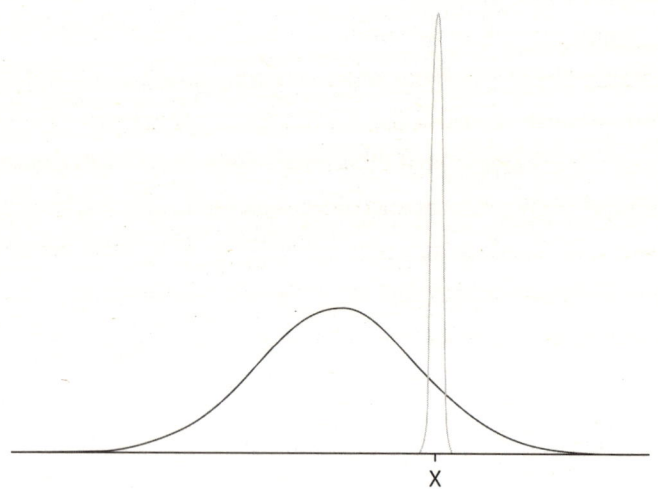

X

Figure 4
If, following a position measurement, a particle is found to be
well localized at point X, its wave packet will then be narrowly
peaked around X as shown, in contrast to the broad hump in Fig. 3.
Conversely, its motion will be very uncertain, spread out over a wide
range of possibilities. Starting with a broad-hump wave packet, the
effect of measurement is to cause the broad wave form to abruptly
'collapse' to the spiky form.

where the particle was prior to the observation, implying that the particle's position has no concrete independent existence.

Let's see how that claim plays out in the case of an electron orbiting the nucleus of an atom. An atom is, of course, very small and the electron moves very fast. Nevertheless, intuition suggests that at any given instant the electron should be somewhere – in front of the nucleus as you view it, perhaps, or maybe to the right. But quantum mechanics says no. There are relative probabilities for these locations, but in the absence of an actual position measurement, it is vacuous to assert that the electron *really is* momentarily at some particular place. Same story for its motion. You may wonder whether at some particular instant the electron is orbiting clockwise or anticlockwise, or vertically or horizontally, or some other orientation, but quantum mechanics asserts that there may be no definite answer without an actual measurement of its motion being made. In fact, for the lowest energy state of the simplest atom (hydrogen) the electron is equally likely to be found at any orientation relative to the nucleus, and equally likely to be orbiting in any plane, and equally likely to be going clockwise or anticlockwise. Evidently, for the ground state of the hydrogen atom at least, it would be wrong to think of the electron as 'going around' the nucleus at all. It's just a stationary fuzzy blob of probability clinging to the nucleus.

The experimenter as creator

Probability goes hand-in-hand with uncertainty. As I just explained, properties like position and motion on the atomic scale are blurry and uncertain, but precisely how uncertain? What is the degree of fuzziness here? An answer was provided by Born's compatriot Werner Heisenberg, who deduced from the properties of the ψ wave that if you know some things well, then you can only know other things badly; it's a trade-off. Take position: if you pin down where a particle is located (see Fig. 4), then you can't know anything about its motion (to be accurate, its momentum). And conversely, if you know the speed, then the particle is represented by a wave extended throughout all space: if the wave is everywhere, the particle can be anywhere. Heisenberg showed that the *combined* fuzziness of both properties (obtained by multiplying the uncertainties of each) can't be less than a definite quantity, depending on Planck's constant (see p. 15). That statement is known as Heisenberg's uncertainty principle, and it's an actual mathematical formula, which applies not only to position and momentum but to other pairs of quantities that are likewise incompatible. Quantum uncertainty doesn't imply anarchy but a precisely legislated degree of wiggle room in the values of physical quantities.

Born's probability rule and Heisenberg's uncertainty principle have obvious implications for cause and effect, and thus for the nature of the physical world. Science was founded on the notion that there are laws of nature that material objects must obey, and obey precisely. If a sharpshooter misses the bullseye, it is put down to a flawed aim; it's not the bullet's fault. We wouldn't accuse it of annoyingly veering off on some new path of its own choosing: the bullet will go where the laws of physics tell it to. Given those laws, once the initial conditions of the gun and bullet are specified, the bullet's trajectory is completely determined. Determinism enables us to predict what will happen when we have complete knowledge of a system; it is the basis of precision ballistics. It was central to science for centuries, until quantum mechanics came along and undermined it. You can aim an electron gun at a target but you take your chances where the electron will actually go. Whatever control you may have over the firing process, quantum uncertainty smudges the outcome. Schrödinger's equation tells you where the wave will go, but nobody can tell you where the particle will go.

It has to be conceded that determinism was always something of an idealization. In daily life, a lot of things that happen seem like a lottery: a drop of rain hits a particular patch of ground, a rolled die shows six, a flipped coin comes down heads. These are examples of unpredictability due to human ignorance. If you knew everything about the coin and its environment, for example, you could (in principle) predict it will land heads up. In other words, the coin isn't defying cause and effect; its outcome is determined (to within quantum uncertainty, which is minuscule for a coin) by causes of some sort. The unpredictability arises because we don't know, or can't possibly keep track of, those myriad tiny causes needed to correctly predict the result. Quantum mechanics, by contrast, is *inherently* indeterministic. Even nature doesn't know in advance what will actually happen moment by moment. The unobserved microworld, it seems, is suspended in a bizarre state of limbo. Which raises the knotty question of what happens when a measurement *is* made.

What happens to a quantum particle when somebody looks?

If someone measures the position of the particle, they can obtain a definite result – a sharp value. Evidently, the very act of measurement *brings into being* a particle-with-a-position, projected into existence from a fuzzy prior state. However, that same someone is free to measure the motion (momentum) of the particle instead, and obtain a definite result for that. In which case, a particle-with-a-speed is conjured up. These two entities – a particle-with-a-position and a particle-with-a-speed – are not the same thing. Both are valid realities after the event, but it's meaningless to try and retrodict what the particle was 'really doing' before. The bottom line is this: *measurements serve to create the physical properties of quantum systems, they do not merely uncover pre-existing realities.*

How surprised should we be that human intervention, such as measuring the position of an atom, creates a new world? After all, we bring about change all the time. If I move a book from one side of my desk to the other, have I not created a new world of sorts? There is a stark difference, however. The book already had a definite position and I merely altered it. In the quantum case, a position measurement doesn't change a pre-existing state of affairs, it *creates* a particle's position from a state of positionlessness, which is far more disturbing, both literally and metaphorically.

It should be clear from the foregoing that a quantum state is fundamentally different from most things we encounter in life (clouds, rocks, trees, people . . .), because the ψ waves of quantum mechanics don't describe *what's there* in the usual sense. Rather, ψ can tell you something about the world, but only statistical information – probabilities and averages – of observable quantities. It says little about a specific case. Nevertheless, it is important to note that the quantum state, if it is specified exactly, contains *everything that can in principle be known* about the system of interest. It is the most complete description one could give. And still it is riddled with uncertainty.

The claim that uncertainty infuses the bedrock of reality was no mere technical detail, but a radical departure from the entire scientific paradigm prior to that point, and it provoked an inevitable backlash. Einstein for one allowed that uncertainty pervaded the atomic realm, but he was convinced there must be a set of concealed causes at work that generated the *appearance* of randomness, after the manner of the flipped coin. If the coin comes down heads, there will be a myriad of tiny forces operating, but too many for us to analyse, so the upshot looks random but actually it isn't. Perhaps quantum uncertainty works like that? Einstein maintained that beneath the quantum world, with its perplexing uncertainty, lies a more familiar realm of definite and precise causes. As we shall see, his challenge led to a decades-long culture war about what is 'really going on' in the quantum world.

Passing through walls

Heisenberg's uncertainty principle says a given particle might be here, or it might be there. But what happens if there is a barrier between here and there? Could a particle be here at one moment and there the next – in effect, passing through the wall, like Harry Potter at King's Cross station?

The short answer is: yes. This was shown by Born's junior collaborator Friedrich Hund in 1927. So long as the barrier isn't infinitely strong, there is a chance that a particle located on one side will later be found on the other side – without breaching the barrier. It's like throwing a brick into a room through a glass window while leaving the pane intact. Known as the tunnel effect, the quantum penetration of barriers has a ready explanation using Schrödinger's wave equation. When the wave encounters the start of a barrier, part of it is reflected, representing the possibility that the impinging particle bounces back off the barrier. But some of the wave creeps inside the barrier, albeit rapidly attenuated (see Fig. 5). How far the wave penetrates depends on the wavelength and the strength of the barrier. For example, if the particle is an electron and the barrier is a repulsive electric field, the penetration depth will depend on how fast the electron is travelling and how strong the electric field is. If the barrier is relatively thin, then a significant fraction of the impinging wave may reach the far side and exit into the force-free region beyond. The amplitude of the exiting wave yields the probability that the particle has tunnelled through the barrier and continues on its way, as opposed to being reflected. The term 'tunnel effect' is a bit misleading, because it may give the impression that the particle batters or drills its way through the barrier, whereas it's more accurate to think of the process as the particle dematerializing on one side of the barrier and re-materializing a moment later on the other.*

An early triumph of the quantum tunnel effect was to

---

* In spite of the somewhat magical feel of the tunnel effect, it has many important real-life applications, such as the scanning tunnelling electron microscope, which can be used to create detailed images of material surfaces at the atomic scale; indeed, it can resolve individual atoms, making it invaluable for studying surface structures and material defects.

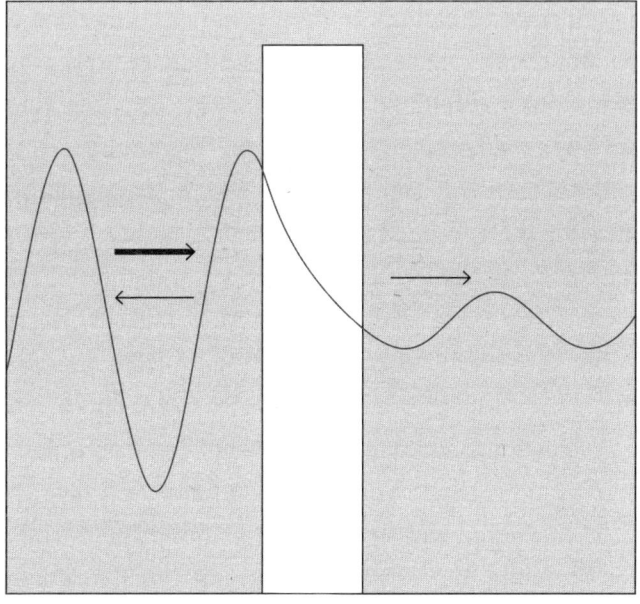

Figure 5

The tunnel effect. A right-moving electron hits a barrier. The quantum wave describing the electron's motion partly reflects back to the left (smaller arrow), the amplitude of the reflected wave representing the probability that the electron will be reflected. But some of the wave enters the barrier, a region that would be forbidden to a classical particle with that impinging energy, where the wave is rapidly attenuated. However, if the barrier is thin, some of the reduced-strength wave will survive to exit on the far side and flow away to the right, representing a probability that the electron will tunnel through the barrier and keep moving rightwards, even though it lacks the energy to surmount the barrier.

provide an explanation of radioactivity (oddity # 6 in Ch. 1). Atomic nuclei are bound together by a strong force that opposes the mutual electric repulsion among the nuclear protons. But the strong force is very short-range, dropping to almost zero just outside the nucleus. The interplay of the two forces – long-range electric repulsion and short-range strong nuclear force attraction – creates a narrow force barrier that is mostly enough to confine the protons. However, in heavy elements like radium and uranium, the binding is a close-run thing. As a result, these nuclei are unstable, and they eventually decay by ejecting an alpha particle (a tightly bound ball consisting of two neutrons and two protons). Puzzlingly though, the energy of an emitted alpha particle isn't enough to actually overcome the force barrier. If an ejected alpha particle hits another identical nucleus, it almost certainly can't get inside; it will just bounce off. In 1928, George Gamow did a calculation to show that the alpha particle escapes its nuclear prison not by jumping over the wall, but by quantum tunnelling through it, and zooming off into the outside world. The quantum nature of the process explains why radioactive elements have a fixed half-life, because there is a certain definite probability that any given nucleus will have decayed this way after a certain time.

Radioactivity is of interest to scientists, but not immediately relevant to daily life. However, there is an example of quantum tunnelling that is literally vital. Life on Earth is sustained by sunshine. The sun is a gigantic thermonuclear reactor in which protons in the core fuse together to make helium nuclei. The temperature in the middle of the sun is about 15 million degrees Celsius, so the protons are banging around at high speed – but not high enough to overcome their mutual electric repulsion and approach within the range of the nuclear force. However, the protons can quantum-mechanically tunnel through the force barrier and fuse together, initiating the first step in what is a rather complicated series of nuclear reactions, without which I would not be writing these words, and you would not be reading them.

Blended alternative realities

Let's dig a little deeper and focus on the notion of a position measurement. Where, oh where could an electron be . . . *now*? It might be here, or it might be there with various probabilities, but as I have explained, that doesn't mean it *is* here or there prior to measurement. It is in some sense *both* here and there – and more generally everywhere: all possibilities co-exist as an amalgam of contending realities. Physicists describe this amalgam by saying that the particle is in a state of *superposition*. Superpositions of different quantum states are a defining feature of quantum mechanics that sits at the foundation of the entire subject. It applies not only to particle positions but to *all* measurable properties: energy, momentum, angular momentum, and so on.

One way to think about superpositions is as *potential* worlds or potential realities. If a particle is in a superposition of many possible locations, for example, it could be described as 'being in many places at once'. The wave might then look like the broad curve shown in Fig. 4. After a rather precise position measurement is performed, detecting the particle to be confined near position X, for example, the wide-humped wave abruptly narrows to something like the spike shown in the figure.*

Following this measurement re-set, the form of the wave going forward will now be the spiky one, not the original extended one. This sudden change in the form of the wave is usually termed the *collapse of the wave function*. The word 'function' here is a mathematical term, meaning a quantity that depends on ('is a function of') other quantities. The wave function is more or less the same as the quantum state, so I shall use the terms interchangeably. For the wave function shown in Fig. 4, the term 'collapse' is rather misleading, as it might be misinterpreted as a vertical flattening rather than a horizontal squeezing. (Sometimes the term 'reduction' is used instead of collapse.) More broadly, the collapse of the wave function is taken to mean the abrupt transformation from a wide range of possibilities to one specific possibility, or at least to a narrower range of possibilities. Importantly, that step is irreversible. Once you've made the measurement, you've disturbed the quantum system and thrown the electron into a new state: you can't go backwards and 'unmeasure' something to recover the initial wave function.

The weirdness of quantum superpositions is well illustrated with an everyday example. It's a perfectly meaningful question to ask whether my tumble dryer spins clockwise or anticlockwise. (Mine spins clockwise when viewed from

---

* Generally, a wave function will be a superposition of an infinite number of components. Mathematically, we can think of such a state as represented by an infinite-dimensional abstract space, technically known as a Hilbert space, after the mathematician David Hilbert. It is not a real space, but a 'possibility space'.

the front. I haven't made a study of it, but it's my impression they all go around that way, for no particular reason that I can fathom. However, my wife tells me that some fancy dryers reverse direction at times.) If I claimed that my dryer spins *both* clockwise and anticlockwise at the same time that would make no sense. Yet that's precisely what can happen in quantum mechanics.

But what does this mean, exactly? If somebody makes a measurement of whether a system is spinning clockwise or anticlockwise, they will always find it to be *either* one *or* the other: never both (obviously). Schrödinger's equation and Born's rule will give you the betting odds for each outcome. It might be 20 per cent clockwise, 80 per cent anticlockwise, depending on the quantum state before the measurement was performed. Or, to take a common example, an electron collides with an atom and bounces off (a process termed scattering). The electron behaves like a wave that ripples out from the impact in *all* directions. (The atom has an associated wave too, but as the atom is much heavier than the electron its own ripples are more confined.) Schrödinger's equation will enable you to predict the actual pattern of waves spreading out from the atom, and Born's rule will tell you how likely it is you will find the electron heading this way or that. When you actually measure where the electron is, you will of course find it to be somewhere. The wave will then suddenly collapse to a definite location.

The measurement serves to re-set the system, following which the wave will begin to spread out again. And now we encounter a curious phenomenon. If a second measurement is made shortly after the first, the wave won't have had a chance to spread out much and so it will collapse again to a position close to the previously determined location. If repeated measurements are made in rapid succession, the poor electron will be frozen in a state of immobility. For example, it might be inhibited from tunnelling through a barrier. This phenomenon is reminiscent of an argument set out by the fifth-century BCE Greek philosopher Zeno of Elea suggesting that all change should be impossible. He used the example of an arrow flying to its target. For the arrow to reach the target, it would first have to pass through the mid-point. But to reach the mid-point, it would need to reach the mid-point of the distance to the mid-point. And so on. If the space between the archer and the target is continuous and infinitely divisible, reasoned Zeno, the arrow would need to visit an infinite number of points before it could reach the target, which would, it might seem, take an infinite amount of time. Hence the arrow could not move. Of course, arrows do move, so this argument became known as Zeno's paradox. It was fully resolved only in the nineteenth century with a rigorous mathematical analysis of the concepts of infinity, continuity and infinitesimals. But repeated quantum measurements provide a modern reprise of the philosopher's argument. Another class of 'paralysis' experiment involves repeatedly measuring the energy of an exited atom. As soon as the atom tries to decay, a measurement snaps it back into the excited state, 're-setting the clock' for decay. This is often called the watched pot effect: it is said that a watched pot never boils, and a watched atom never decays.

Acts of measurement provide us only with restricted snapshots of the vast quantum realm of possibilities – that wonderland of almost limitless potential with which I began my account (p. 3). We humans can never directly observe this wonderland. Every measurement, every observation, collapses the wave function onto just one of the (maybe infinite) possibilities – an infinitesimal sliver of reality. Left unobserved, however, nature restlessly twists and turns the wave function of the undisturbed system, cavorting around the boundless combinations of states, an elaborate ballet, not of objects, but of potential realities.

CHAPTER 3

What Lies Beneath

'Do you really believe the moon isn't there when nobody looks?' Einstein once asked.[1] The point he was trying to make is that if unobserved atoms are unreal, and the moon is made of atoms, then the unobserved moon should also be unreal. I'm stating this bluntly: quantum mechanics doesn't actually say that an atom – or any quantum particle – has *no* existence until it is observed. It's just that quantum objects aren't real in the common-sense definition of that term, which assumes that material bodies possess properties like position, motion, shape, energy, rotation and so on. By contrast, quantum particles manifest their properties only when a specific measurement is made. It is better to think of an unobserved, or yet-to-be-observed, atom as less-than-common-sense-real rather than out-and-out unreal. Nevertheless, Einstein's point isn't lost. Whatever ghostly entity a less-than-real atom might be, then it seems as if the moon, made of many such insubstantial entities, shares their less-than-realness. So, if nobody is looking at the moon, suggested Einstein, it should be hovering in a vague state of incipient, unfocused existence. Which seemed, to him, self-evidently absurd.

The account of quantum mechanics I have given so far is what we might call the textbook version or the party line.

But Einstein was never happy with that 'official account' and the indistinct picture it paints of physical reality. He was not the only dissenter, but he was perhaps the most persistent. 'I have become imbued with great fear of the "evil spirit" of quantum mechanics, thinking that I was burying my head in the sand like an ostrich,' he wrote in a letter to a colleague.[2] As I mentioned in the previous chapter, Einstein clung to the belief that beneath the quantum domain, with its weirdness and uncertainty, there lies a more familiar world of definite deterministic causes. This mundane explanation became known, for obvious reasons, as the *hidden variables* theory. What was lacking, however, was any clear evidence that those all-important hidden variables actually exist. If there really are physical processes operating beneath the quantum world, shouldn't their clandestine activity be detectable somehow?

It took several decades to design a definitive test of Einstein's ideas. To explain what was involved, let me start with something more basic – a famous experiment first performed over two centuries ago, one that the renowned physicist Richard Feynman regarded as encapsulating all 'the basic peculiarities of all quantum mechanics.'[3]

Wave or particle?

Long before the advent of quantum theory, there was a protracted debate about whether light is a wave or a stream of particles. Isaac Newton, for example, suggested that light is made up of 'corpuscles', whereas his contemporary, Christiaan Huygens, explained many features of light by treating it as a wave. Both theories had their merits. How to settle the matter? A decisive experiment had to await the attentions of a talented eighteenth-century polymath named Thomas Young. Young trained in Edinburgh as a medical doctor and went on to study colour vision. He also became famous for deciphering Egyptian hieroglyphs. But it was for his 1801 experiment with light that he is best remembered. Young devised a simple piece of apparatus to demonstrate for light what is familiar from water waves: the phenomenon of wave interference. Drop two stones side by side onto the surface of a smooth pond and watch the ripples spread out and overlap to form a distinctive pattern (see Fig. 6, where the white circles represent the stones). Where the peak of one wave meets the trough of the other, they cancel, producing a flat patch; this is called *destructive* interference. However, where peak meets peak and trough meets trough, the waves reinforce to make a bigger hump and a deeper pit; this is *constructive* interference.

Young wanted to do the optical equivalent of dropping two stones in a pond together, and he devised a clever way to do it, shown schematically in Fig. 7. A compact light source illuminates an opaque screen A containing two slits. Light passes through the slits, spreads out a bit (a phenomenon called diffraction) and makes an image on a second screen, B. If light consisted of particles (corpuscles), the image would be two bright vertical bands side-by-side where the light emanating from each aperture struck the screen B. But if light is made of waves, then the two slits would be like the two stones, and waves rippling from both slits would overlap to form *many* bands – an interference pattern – bright where the two waves arrive in step and reinforce, dark where they arrive out of step and cancel, as shown in the figure. Sure enough, Young observed these interference bands (usually

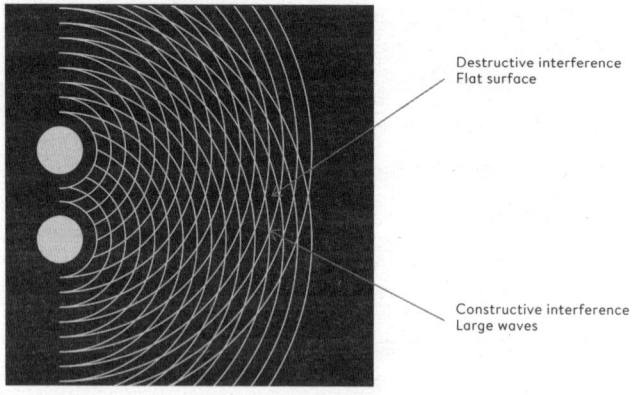

Destructive interference
Flat surface

Constructive interference
Large waves

Figure 6
Water waves emanating from two nearby sources overlap to make
a distinctive pattern known as interference.

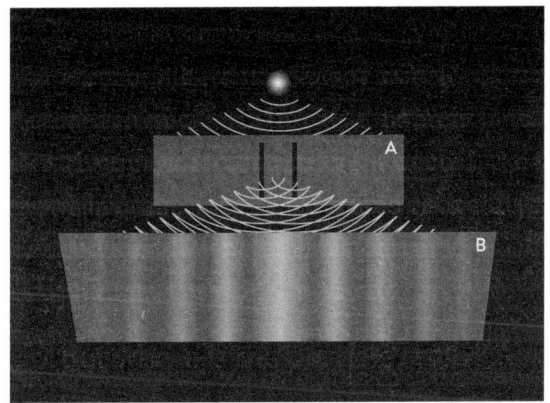

Figure 7

Two-slit experiment. Light from a point source falls on a screen with two narrow nearby slits in it (A). The light passes through the slits to an image screen (B), where it creates a pattern of bright bands called interference fringes. This pioneering experiment proves that light consists of waves.

called fringes) proving that light is a wave. The fringes are tricky to produce because the wavelength of visible light is so small – about 100 nanometres – but they are clearly visible.

The lesson of Young's pioneering experiment was in the minds of quantum physicists grappling with the wave nature of matter in the 1920s. For example, could one do a Young's-type interference experiment for matter waves too? The answer is yes, but it's technically more challenging, and it actually took until 1961 for the job to be done properly, by Claus Jönsson of the University of Tübingen.[4] The apparatus is depicted schematically in Fig. 8(a), showing a type of gun spitting out a stream of electrons that passes through a pair of slits. The particles are then detected individually on a screen that emits a little flash of light when an electron strikes it. Fig. 8(b) shows the results. Each white dot records the arrival of a single electron, and the mottled pattern is produced by the steady accumulation of dots from many such transits. Although Heisenberg's uncertainty principle forbids anyone knowing exactly where each individual electron will hit the screen, the accumulated results of many such transits clearly displays the repeating band structure of interference fringes. Bingo! Electrons behave like waves.*

All of which leaves us with that nagging question: how can something be both a wave and a particle at the same time? This is where we start to part company with common sense and everyday intuition. What if, in the two-slit interference

---

* Since that pioneering effort, matter-wave interference has also been demonstrated with whole atoms and even with molecules as large as carbon 60 – the well-known 'buckyball'. Devices that use quantum interference effects for commercial and scientific purposes are called interferometers, and I shall later discuss several examples.

(a)

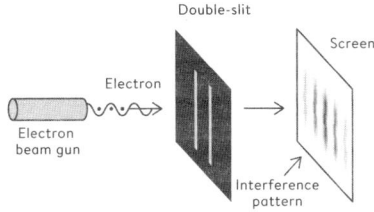

(b)

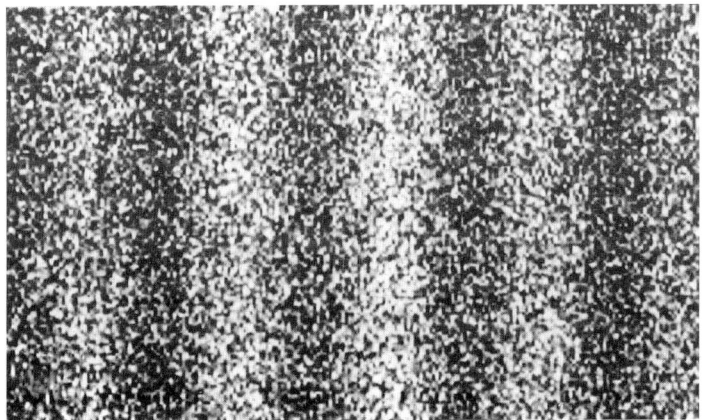

Figure 8
A two-slit experiment performed with electrons (a). Interference fringes are observed in this case too (b). Each spot records the arrival of one electron; the speckled pattern builds up and gets more pronounced as more and more electrons are detected.

experiment, the electrons are fired *one at a time*? Because each electron makes a little dot when it hits the image screen, it manifests itself at that stage as a particle, localized at one spot (literally, in fact). Does that mean it was a particle *before* it was detected, when it was traversing the slits? If so, the electron must obviously go through *either* one slit *or* the other, but not both, because an electron is a little speck of matter. On the other hand, if the experiment is performed with just a single slit, no interference pattern is observed when many dots are accumulated on the screen. Both slits are needed to make the distinctive pattern; yet any given particle could travel through only one slit.

You might be thinking that a mischievous experimenter could station a detector near each slit and when one of them signalled 'particle heading my way', the experimenter could quickly block the other slit. Surely it would make no difference to the outcome? When this argument was mooted in the 1920s, the possibility of performing such a practical experiment with a single quantum particle was regarded as fanciful. Today, however, it is possible to do more or less what I just described. And the result is astonishing. The very act of finding out 'which slit' serves to disrupt the experiment and to destroy the interference pattern. The wavelike pattern survives only if the path taken by each electron remains ambiguous, i.e. if the experimenter couldn't tell which slit that electron goes through. There has to be at least the *possibility* of any given electron going through either slit for the interference pattern to emerge.

As if this weren't baffling enough, things get even stranger. The disruption of the pattern isn't simply due to the detector at the slits haphazardly knocking each particle off its trajectory as a result of the detection. It turns out that you can determine the slit a particle traverses without ever getting near it. This refinement is best discussed when the experiment is performed with photons rather than electrons, because a single photon can be turned into two less energetic photons by passing it through a special type of crystal. When that is done, the offspring fly out in different directions. By placing the crystal just after the slits, you can use one progeny photon for the interference experiment and the other (by now far away) photon to figure out which slit the original photon went through. It is measurement by proxy. If you manage to acquire this information for each photon that traverses the slit screen, you discover that no interference pattern forms. Evidently, the mere *knowledge* of what a photon is doing is enough to alter the experimental results, even without directly touching the photons concerned.

Completing the story

Waves and particles, superpositions and uncertainty, multi-faceted reality – it was all bewildering to the founders of quantum mechanics as they struggled to shoehorn the theory into a comprehensible framework. By 1927, results and applications were coming in thick and fast, but there was no agreement on what to make of it all. In October of that year, the Fifth Solvay Conference was held in Brussels. The Belgian industrialist Ernest Solvay had founded these conferences in 1911 with the express intention of addressing foundational, unsolved issues in physical science. The 1927 meeting, entitled 'Electrons and Photons', featured an all-star line-up of quantum afficionados: Bohr, Born, de Broglie, Einstein, Heisenberg, Wolfgang Pauli and Schrödinger. This get-together provided a chance for the physics community to finally sort out quantum mechanics and resolve the paradoxes. Einstein was the curmudgeonly sceptic, continually trying to identify a fatal flaw or contradiction in the theory.

The person who emerged as the elder statesman of this esoteric coterie was the Danish physicist Niels Bohr. Through his influence, there emerged a rough consensus on how to think about quantum mechanics, which became known as the 'Copenhagen interpretation' – which I have thus far called the 'standard' or 'textbook' interpretation – though not all aspects of it were necessarily endorsed by Bohr himself. The central dictum of the Copenhagen interpretation is that reality is anchored in actual measurements of observable quantities, and questions about the shadowy goings-on in between measurements are mostly meaningless. In later years, Bohr's collaborator John Wheeler put it like this, in typical poetic style: the quantum world, he said, is like 'a great smoky dragon'. The dragon has sharp teeth, and a sharp tail, corresponding to concrete measurements at two places and times, but in between is a cloud of uncertainty. Trying to assign an independent existence to the physical world between measurements is a fool's errand.

The Copenhagen interpretation remained vague, however, on a fundamental issue: what, exactly, constituted 'a measurement'. According to Bohr, a crucial aspect of a quantum measurement is that the result must be sufficiently amplified that it is discernible on a macroscopic scale, where it can be inspected by a human being and communicated to another in plain language. For example: 'the Geiger counter clicked, so we can be sure that the atom has decayed,' or 'the meter on the detector shows that the electron has tunnelled through the barrier and emerged moving to the right.'

Here is the point made in Bohr's own words:

> It is decisive to recognize that, *however far the phenomena transcend the scope of classical physical explanation, the account of all evidence must be expressed in classical terms.* The argument is simply that by the word 'experiment' we refer to a situation where we can tell to others what we have done and what we have learned and that, therefore, the account of the experimental arrangement and of the results of the observations must be expressed in unambiguous language with suitable application of the terminology of classical physics.[5] [Emphasis in original.]

Bohr's reference here to 'unambiguous language' is somewhat ironic because he had a reputation among his peers for talking in riddles and ambiguities, often while pacing round and round, collecting his thoughts. I once discussed the atmosphere in pre-war Copenhagen with the physicist Carl Friedrich von Weizsäcker at a conference for Bohr's centenary. He said that young physicists like him would hang on Bohr's every word, but sometimes, just as the great man seemed to be crystallizing a thought, his voice would drop to a whisper and his audience would strain to catch the all-important conclusion. 'It was almost as if he didn't want us to hear,' mused von Weizsäcker. And perhaps there was a method in Bohr's obfuscation, for it was part of his philosophy that ambiguity and apparent contradiction pervade the world. Perhaps, by leaving his audience befuddled, he might provoke them into spawning a new idea or concept.

The cat paradox

In spite of Bohr's influence, many physicists believed – and continue to believe – that the Copenhagen interpretation is a fudge, an excuse to ignore the knotty problem of how to marry the shadowy quantum realm to the concrete experience of daily life. To ram home the absurd conclusions that stem from dismissing this problem, Schrödinger came up with a famous thought experiment.[6] In a letter to Einstein, he envisaged a cat incarcerated in a steel chamber with a small quantity of radioactive substance and a Geiger counter. A contraption ensures that, if the counter detects a radioactive emission – a fundamentally quantum process – it operates a mechanism that smashes a phial of cyanide, whereupon the cat is rapidly dispatched (see Fig. 9). After a certain period of time, there will be a definite probability that a decay event has occurred and the cat is dead. But there will also be a probability that the decay hasn't occurred and the cat is alive. In the absence of an actual observation, quantum mechanics would seem to say that the cat is in a hybrid live–dead state – a superposition of living and dead, to express it technically. The cat 'paradox' as it came to be known forcefully demonstrates the oddity of quantum reality. An atom, shadowy thing that it is, is one thing, but a cat?

Schrödinger's cat experiment throws into stark relief the problem of where, on the chain from atom to cat to physicist, ghostly superpositions give way to solid reality: in other words, the transition from this-and-that to this-*or*-that. 'Where is the cut to be between the description by the wave function and the classical description?' asked Schrödinger's

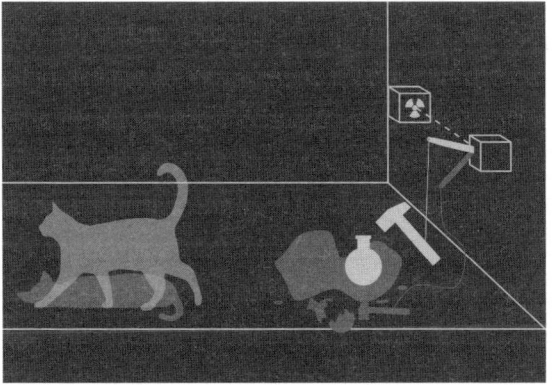

Figure 9

Schrödinger's cat. In this (entirely thought) experiment, the cat in the box is dead or alive depending on whether a radioactive atom has decayed or not. Quantum mechanics applied to the entire contents of the box describes a seemingly paradoxical superposition state in which the cat is both dead and alive at the same time.

contemporary, Heisenberg.[7] One response to the paradox is to say that there is *no* point on the upward chain from atom to cat at which something fundamentally new intrudes. Maybe it's a case of: the cat is dead but with a tiny smidgeon of 'aliveness' in it, or alive with a smidgeon of deadness. (Whatever that means.) But of course, that still leaves us with a dichotomy – which of those two is it? Can we know without looking? And if we don't know, is it still meaningful to say it is one or the other? And what about the cat itself? Can it not observe its own aliveness? Does it really need a human to decide the matter? The answers to all such questions were left hanging at the time. Much later, Edwin Jaynes, a renowned expert on probability theory, expressed his dissatisfaction with the unfinished state of the subject. The standard treatment of quantum mechanics, he observed, 'is a peculiar mixture describing in part realities of Nature, in part incomplete human information about Nature – all scrambled up by Heisenberg and Bohr into an omelette that nobody has seen how to unscramble . . . if we cannot separate the subjective and objective aspects of the formalism, we cannot know what we are talking about; it is just that simple.'[8]*

The Copenhagen interpretation places the 'reality onus' squarely on acts of measurement. It's at the point of

---

* The cat experiment, I hasten to add, was never intended to be carried out on real cats: it is entirely hypothetical, though it does make for lively theatre. I once presented a television documentary in which I rashly offered to symbolically enact Schrödinger's thought experiment in the studio. The subject chosen for this little charade was a studio moggie known for its extreme docility. The good-natured animal tolerated being stuffed in a box a couple of times, but by the third retake it was becoming decidedly neurotic and claustrophobic, meowing and clawing the sides of the box and totally giving the game away: the cat was demonstrably alive in there.

measurement that the all-important, but deeply puzzling, reality-projecting, dice-playing part of quantum mechanics kicks in. But the true weirdness of quantum mechanics is that the experimenter is free to choose *which* quantity to measure, and hence which 'reality' to bring into being; for example, an-atom-with-a-speed or an-atom-at-a-place. Bohr expressed it by saying that the particle has two 'complementary' observables – position and momentum – but they cannot both be manifested at the same time.

So, we are left with a conundrum, referred to as the *measurement problem*. According to the Copenhagen interpretation, at the level of atoms, the system is described by the quantum wave function ψ – an abstract mathematical object that doesn't itself correspond to an observable physical entity. But at the level of the apparatus, and the experimenter (or the cat), the description is classical (i.e. non-quantum). So where is Heisenberg's cut located? Quantum mechanics is silent on the subject: it doesn't tell us where that cut is situated. But here is the point: it doesn't seem to matter, because quantum mechanics predicts the same experimental results anyway. Consequently, most physicists choose to ignore the measurement problem altogether and just get on with the job. This 'shut-up-and-calculate' school advocates simply using the mathematical theory of quantum mechanics – Schrödinger's equation and the Born rule – to calculate things of interest and test the predictions with experiment, and to stop fretting about what is really going on. And it works brilliantly! Molecular shapes and motions, atomic collisions, forces between atoms, nuclear structure and fission, the electrical and thermal properties of solids, the nuclear reactions inside stars, the behaviour of light – all these things tumble out of the theory, as does the entire field of subatomic particle physics. By the outbreak of the Second World War, most of the essential properties of matter were well understood within the framework of quantum mechanics, without any reference to the measurement problem.

Quantum technology 1.0

It didn't take long after Schrödinger published his famous equation before scientists and engineers spotted some promising technological opportunities. An early invention was the electron microscope. This instrument can resolve tiny structures such as viruses that are too small for light microscopes; in fact, they can now image individual atoms. Another handy device developed in the 1930s was the photomultiplier tube, which can detect and amplify very faint light signals. Today's photomultipliers can detect single photons. A milestone in quantum engineering was the transistor, invented in 1947 by John Bardeen, William Shockley and Walter Brattain of Stanford University. Exploiting the quantum properties of silicon semiconductors, transistors became critical building blocks for much of modern electronics, eventually becoming miniaturized and embedded in the familiar microchip. The silicon transistor in turn gave birth to Silicon Valley – the environs of Stanford University – now the host to a large number of high-tech industries that are in many ways downstream beneficiaries of this early invention.

Then came the laser, an acronym for Light Amplification by Stimulated Emission of Radiation, first demonstrated in 1960. Initially it was regarded as a mere novelty – 'a solution looking for a problem', in the words of its co-inventor, Charles Townes, because there were no obvious practical applications apparent at the time. Today, of course, lasers have found many important uses in communications, medicine, industry and entertainment, not to mention supermarket check-out machines. Taken together, the transistor and the laser laid the foundations for the entire information age, with long-range optoelectronic communications, the internet and mobile phones. We live in a connected world largely thanks to quantum principles at work in countless devices we take for granted in daily life. And, of course, our political world order has been shaped and re-shaped by nuclear power and nuclear weapons, also very much a product of the quantum age: a nuclear explosion is a direct manifestation of quantum mechanics applied to nuclear structure and reactions.

These well-known examples have merely scratched the surface. Quantum technology penetrated many sectors of industry throughout the twentieth century – I have listed some of the more familiar practical applications in Table 1. In one way or another, they all use the distinctive properties of quantum systems, such as the wave nature of matter, superpositions and the tunnel effect, which emerged in the early 'gold rush' days of quantum physics.

I call these products Quantum Technology 1.0. They were mostly designed by 'shut-up-and-calculate' researchers. To be sure, the baffling weirdness was always there under the hood, but of interest only to philosophers and a small group of nit-picking theoretical physicists. Even today, although the deep nature of the theory remains ill-understood, most scientists and engineers continue to just forge ahead, accepting

Table 1: Quantum Technology 1.0

Date	Quantum Effect	Eventual Products
1933	Wave nature of electrons	Electron microscope
1933	Quantum tunnelling measured	Many electronic devices including USB flash drives (memory sticks)
1938	Nuclear magnetic resonance discovered	MRI machines
1938	Nuclear chain reaction created	Nuclear power, nuclear weapons
1938	Superfluid first produced	Ultra-sensitive thermometers, cryogenic devices
1946	Microwave amplification by the stimulated emission of radiation (MASER) observed	Lasers
1947	Transistor invented	Much of modern electronics and computing
1955	First atomic clock invented	Highly accurate time-keeping, the GPS navigation system
1957	Superconductivity explained by quantum processes	Superconducting magnets, high-sensitivity detectors, maglev trains
1981	Quantum dots invented	Improved LED displays, solar cells and medical imaging

quantum mechanics for what it is and applying it to their systems of interest. You don't need to worry about the nature of reality to design a lucrative new microchip or laser that performs as predicted. A dollar is a dollar, even if the electrons and photons that earn it are shadowy and insubstantial.

But a second quantum revolution – Quantum 2.0 – is underway. What sets it apart from the first is that the weirdness is not merely confronted head-on: it is actually regarded as a resource to be exploited. A puzzle has been turned into a technological opportunity. Part of the reason for this quantum renaissance is that physicists and engineers can now control single electrons, atoms and photons, and can manipulate properties such as superpositions with great precision. It's also possible to encode information in individual quantum states in a radically new way. This is the exciting field of 'Reality Engineering', in which customized wave functions are sculpted and controlled for practical ends. To understand how it all works, we must look under that quantum hood and understand how phenomena that a century ago were mostly dismissed as simply weird, if not nonsensical, began to be analysed and actually harnessed, forming the basis for a new technological age.

CHAPTER 4
Reality Wars

In science, it is often the case that major revolutions are born from arcane academic disputes about foundational concepts. The Industrial Revolution stemmed from a debate among physicists over whether heat is a type of invisible fluid or a product of mechanical activity on a microscopic scale; clarifying the relationship between motion and heat led to the laws of thermodynamics and the invention of the steam engine. The digital computer emerged from attempts in the 1930s by mathematicians and logicians to resolve the paradoxes of infinity and self-reference. The historical roots of Quantum 2.0 can similarly be traced back to the 'reality wars' of the 1920s and 30s, which at the time were little more than philosophical quibbles based on imaginary experiments, along the lines of 'what would happen if one could in principle do such-and-such and measure this-or-that?' By the 1970s, technology had advanced enough for some of the early thought experiments to become real experiments. Although the primary purpose was to resolve the simmering disputes about the nature of quantum reality, it transpired that these very experiments provided the critical enabling technology to propel quantum mechanics into a new and glittering age. To grasp these technical advances, we need to look back in history to the original thought experiment, concocted in 1935 by the mischievous and ever-ingenious Albert Einstein.

Einstein's spooky thought experiment

Einstein's stubborn opposition to quantum mechanics, at least in its Copenhagen interpretation, eventually spilled over into the public domain. 'Einstein Attacks Quantum Theory' announced *The New York Times* on 4 May 1935. The subject of the article was a paper entitled 'Can Quantum-Mechanical Description of Physical Reality Be Considered Complete?', in which Einstein and co-authors Nathan Rosen and Boris Podolsky developed a clever argument that something must be missing from the quantum description of the world.[1]

The EPR thought experiment, as it came to be called, goes something like this. Suppose two identical particles come together and then fly apart symmetrically in opposite directions, as in Fig. 10(a). Quantum mechanics can assign a single state (i.e. a wave function) that describes this two-particle system jointly. Suppose that moments later a physicist – call her Alice – decides to measure the position of one of the particles. Prior to measurement, the result is uncertain, in accordance with Heisenberg's principle. Consider that Alice finds the

Figure 10
(a) The Einstein-Podolsky-Rosen (EPR) thought experiment. Two particles interact then fly apart, retaining a 'spooky' link via their joint quantum state. Experimenters Alice and Bob have measuring devices (black boxes) and carry out separate measurements on their respective particles.
(b) A source atom emits a pair of entangled polarized photons with unknown but parallel polarization direction. Alice and Bob have polarizing crystals which may or may not allow their respective

(a)

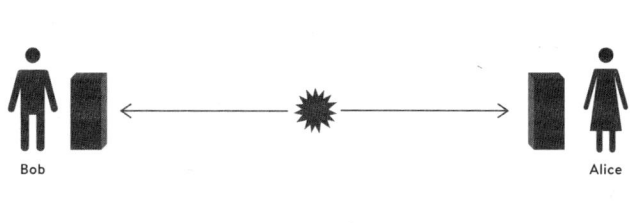

(b)

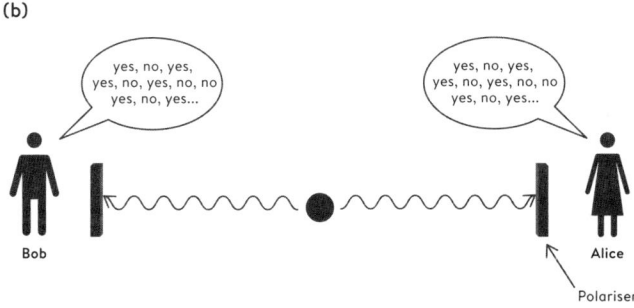

(c)

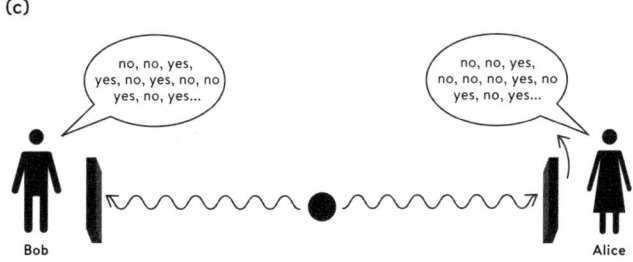

photons to pass through. If their crystals are oriented identically, they will always get the same result, yes or no, over many trials. There is 100 per cent correlation.

(c) Same set-up as in (b), but now Alice has rotated her polarizer through some angle relative to Bob's. The couple no longer get the some results every time: there is reduced correlation, by an amount depending on the angle chosen. Quantum mechanics and hidden variables (pre-existing reality) predict *different* amounts of reduction. Laboratory experiments confirm quantum mechanics.

particle is, say, 10 metres to the right of the collision point. Then, by symmetry, she can infer that the other particle must be 10 metres to the left of the collision point, without having to actually perform any measurement on the remote particle. And, of course, this can be verified if an accomplice – call him Bob – simultaneously makes a position measurement on the other particle, and the couple then compare notes. But now what happens if Alice decides instead to measure, not the position, but the speed (momentum) of the first particle? Once again, that quantity is uncertain prior to measurement, but the measurement yields a definite result: 1,000 metres per second, for example. Again, by symmetry, Alice immediately knows that Bob's particle must also be moving at 1,000 metres per second in the opposite direction. And in this case too, Bob could confirm Alice's inference by conducting his own momentum measurement. It looks as if between them Alice and Bob know *both* the positions and speeds of the particles at the same time, in contradiction with the uncertainty principle.[2]

The real sting in the experiment is that the act of measuring her particle *immediately* tells Alice about Bob's particle too. If Alice is a bit closer to the mid-point, she could know the result *before* Bob makes his measurement. And since she can choose which property to measure – position or speed – Bob's particle must surely 'be ready' with *both* quantities fully determined before his measurement is made, in order for him to get 'the right answer' consistent with Alice's result. But according to standard quantum mechanics, particles don't actually possess well-defined properties prior to measurement. Alice's particle springs into well-defined existence only at the moment it is measured – either as a particle-with-a-position, or alternatively as a particle-with-a-speed, depending on which property Alice (perhaps whimsically, at the last moment) chooses to measure. Thus, there is a flagrant contradiction. Einstein and his colleagues concluded that quantum mechanics must therefore be incomplete, because if particles have potentially knowable properties ahead of measurement, they must possess some 'elements of physical reality' – their words – not captured by quantum mechanics.

How could the particles be in cahoots? One obvious answer would be: if the result of Alice's measurement instantly told Bob's particle what to do. ('She's measured the momentum, so be ready with that answer!') Maybe there is an unknown long-range force that operates between Alice's equipment and Bob's to ensure consistency of results? Notice that the EPR argument doesn't depend on how far apart Alice and Bob are – they could be on opposite sides of the galaxy – so this would have to be a *very* long-range force. Trouble is, the conspiratorial influence passing between Alice's equipment and Bob's equipment needs to be instantaneous. But according to Einstein's theory of relativity, no physical effects can go faster than light, so Einstein absolutely didn't like *that* idea (and nor do most physicists).

Could it be the case that the measurements are consistent because the two particles cooperate from the start? When they flee their common point of origin, they might carry some sort of imprint telling them how to behave in the event of a measurement of this or that quantity. An unknown mechanism – some sort of hidden variables – could ensure that whatever got stamped on Alice's particle was the same as that on Bob's particle, so Alice and Bob's measurements would always concur.

This is not far-fetched. Correlation without causation, stemming from a common point of origin, is familiar in daily life. For example, if you arrive at your ski resort to discover you have accidentally packed only your left-hand glove, you can be immediately sure that the one you left at home is a right-hand glove. Trouble is, this explanation assumes that the particles possess definite predetermined properties from the get-go, prior to measurement, in contradiction with the standard Copenhagen interpretation of quantum mechanics, which holds that quantum particles manifest their properties only when a specific measurement is made. But if pre-existing reality is rejected, it looks like some sort of telepathic quantum link is at work, an invisible thread stretching across space. Einstein disparaged that notion, calling it 'spooky action at a distance'.[3]

The EPR argument threw Bohr, the leading proponent of the Copenhagen interpretation, into a tizzy. His response was characteristically rambling and vague, and dwelt on how far-apart measurements could influence each other, not physically, but in some sense contextually. Much confusion was caused by each physicist's choice of words. Over the next few years, both Einstein and Bohr shifted positions as they sparred with each other on this issue, each striving to make sense of quantum weirdness from their contrasting perspectives. Einstein did not, however, waver from his hidden variables theory – though he didn't attempt to develop it fully, either.

The matter remained inconclusive when the Second World War intervened, pausing this apparently abstruse debate. Then, in the early post-war years, a physicist named David Bohm began to put flesh on the bones of Einstein's hidden variables theory. Bohm was an American-born son of Jewish immigrants. He worked with Einstein at Princeton, and was recruited onto the Manhattan bomb project but didn't get security clearance because of his flirtation with communism in his youth. (Same as Robert Oppenheimer, who nonetheless did get clearance.) After the war, Bohm fell victim to the McCarthy witch-hunt and was hounded out of Princeton, in spite of Einstein's support. Decamping to Brazil, then eventually to London, he continued his work on trying to reinstate a type of determinism (see p. 28) in quantum mechanics by incorporating both waves and particles together. The basic idea is that the waves, which change and move in accordance with Schrödinger's equation, constitute the hidden variables that serve to guide the particle, a bit like a surfer riding an ocean wave. Drawing upon ideas of de Broglie (who first introduced the notion of matter waves in 1924), Bohm formulated a mathematical theory in 1952 that seemed to reproduce the results of quantum mechanics within a completely deterministic framework. The apparent randomness of quantum uncertainty, it argues, arises because of a lack of knowledge about the initial conditions of the waves and the exact positions of the particles. Although not many physicists went along with his theory (and still don't), Bohm attracted a large following and became something of a public intellectual in the 1960s.

Around this time, I was a PhD student in London. I went to meet him, to discuss a calculation I was stuck on. I was rather nervous – a callow student seeking an audience with a scientific superstar. But Bohm was very gracious. We had a long discussion about a commonly used randomness assumption in quantum field theory that I needed to justify in my calculation. A remark of his that sticks with me is that progress in science is often made by dropping assumptions rather than slavishly adhering to them. I've often wondered over the years what concepts and ideas we take for granted in fundamental physics might profitably be questioned.

Box 1

Polarization of light

Light waves are oscillations of the electromagnetic field. The direction of oscillation, which is perpendicular to the direction of motion, is called the *polarization* of light. To envisage it, hold your hand out with your index finger pointing forward, like the barrel of a pistol. Hold your thumb vertically. Imagine a beam of light emanating from your finger. If the light is polarized vertically, your thumb indicates the direction in which the electric field oscillates. Now rotate your wrist, keeping your finger pointing forward. Your thumb now points in the direction of oscillation of another polarization angle. A disorganized beam of light will be a random superposition of many possible polarization angles, but when light passes through certain polarizing crystals it can emerge with just a single polarization direction, the actual angle depending on the orientation of the crystal. (This phenomenon is utilized in polarizing sunglasses.) If a beam of polarized light encounters a second crystal rotated at right angles to the first, its passage will be blocked; no light will get through the second crystal. If the crystals are parallel, all the light will get through. At intermediate relative orientations of the two crystals, a given photon will be transmitted with a specific probability less than one, depending on the angle. That 'maybe transmit, maybe don't' property holds the key to the definitive test of quantum reality.

Tangled reality

For a long time, the EPR thought experiment was relegated to a backwater of physics. Does a particle really have properties before measurement or not? Who cares? So long as it makes no practical difference, the question is akin to asking whether a tree makes a noise when it falls in an uninhabited forest. But this state of affairs changed dramatically thirty years after the EPR paper was first published. In the mid-1960s, it was discovered that what was seemingly just an obscure inter-pretational dispute does indeed make a difference and, as it happens, it is that very difference that lies at the heart of the great quantum information revolution taking place today.

The 'spookiness' implicit in the EPR experiment derives from the special joint quantum state of the two particles, where-in each particle 'knows something' about the other, even when far apart. There is a large class of such states, and the particles in question are referred to as being *entangled*. Entanglement has no classical counterpart, but it occupies a central place in quantum physics; Schrödinger described it as 'the characteristic trait of quantum mechanics'.[4] From the point of view of a real, as opposed to a thought, experiment to address Einstein's challenge, the best set-up is to use entangled photons. Some atoms obligingly emit photons in entangled pairs with parallel polarization directions (see Box 1), so that measuring the polarization of one instantly tells you about the polarization of the other.

Suppose one photon goes to Alice, the other to Bob – our imaginary assistants – who can then carry out polarization measurements on them separately, as in Fig. 10(b). They can place a polarizing crystal in the path of their respective photons and record whether their photon passes through it or not. If you take Einstein's position, the photons will be emitted from the atom with the same definite polarization direction already imprinted on them by some unknown hidden variables, so that when a photon hits the polarizer the probability of it going through will depend on the angle between the photon's pre-existing polarization direction and the orientation of the polarizer. Atoms spit out photons with random polarizations, so for each entangled pair it will be different from the last, which means sometimes the photon will go through, sometimes it won't. Alice and Bob can go ahead and record their respective results as the photons arrive one by one: yes, no, yes, yes, no . . . and then sit down together to compare results. If they choose to orient their respective polarizers at the *same* angle, they will always get the same result, i.e. 100 per cent correlation. That's because, in the hidden variables interpretation, the photon pairs would arrive at the polarizers carrying identical polarization imprints, so Alice and Bob would be performing identical experiments. But it turns out that standard quantum mechanics *also* predicts 100 per cent correlation. Therefore, the experiment as described so far cannot discriminate between the two rival theories.

Now we reach the critical refinement: Alice could choose one orientation for her polarizer and Bob another. For example, if they orient their polarizers perpendicular to each other, they will always get perfect *anticorrelation* (no for one, yes for the other). In that case too, both hidden variables and standard quantum mechanics agree. However, when the polarizers are set *obliquely* to each other – as in Fig. 10(c) – there will be some intermediate degree of correlation (e.g. Alice yes, Bob no; Alice no, Bob no; Alice yes, Bob yes; Alice no, Bob yes . . .). If the photons did carry matching imprints from the point of origin, the imprinting mechanism couldn't know in advance what angles Alice and Bob might choose: our doughty assistants could select those angles *after* the photons have left the source. Suppose Alice and Bob pick random angles for their polarizing crystals. When the photons hit the crystals, the angles between their polarization directions and the crystal orientation will also be random, and it's then a matter of simple statistics to figure out how often on average pairs of entangled photons will go through both crystals or neither (see Box 2). But – and this was the big surprise – quantum mechanics predicts a *different* average coincidence rate from the hidden variables theory. That is, according to standard quantum mechanics, even if Alice and Bob select their angles after the photons have set out, they should get matching results *more often* than would be the case if the photon pairs began their journey with definite polarization directions imprinted on them in advance of hitting the crystals.[5] So it does make a difference after all whether particles possess fixed elements of reality ahead of being measured. It's not just a matter of philosophy.

All this was enshrined in a celebrated mathematical theorem by theoretical physicist John Bell, who proved that, for certain entangled quantum states, the results of widely separated measurements are correlated in a manner that cannot be reproduced by any hidden variable theory that doesn't break the light barrier.[6] Bell's theorem, in the form of an inequality, was published in 1964, and represents an iconic result in the history of physics. Bell himself became something of a celebrity, so profound were the implications of his theorem.

In spite of his fame, Bell was a gentle and unassuming Irishman from Belfast and shunned the limelight. He worked mostly at CERN, the European nuclear research laboratory near Geneva. He kindly invited me to visit and give some lectures there in the 1980s, and during our several conversations I asked him how long he had taken to come up with his famous theorem. He thought for a moment and replied, 'One weekend.' Apparently, he had been to a lecture on hidden variable theories a week earlier and had been quietly mulling over the subject, but the clinching mathematical argument hadn't taken long to fall into place.

Although Bell's theorem was a landmark in science, it was still just a mathematical exercise about a thought experiment. To find out for sure whether beneath the weird world of quantum mechanics lies a more common-sense reality, someone needed to do a real experiment, and with luck, discover whether nature sided with Einstein, or with quantum mechanics.

Box 2

Bell's breakfast bar

To get the drift of Bell's reasoning, imagine a breakfast bar that offers eggs, bacon and tomatoes. Some people take all three, some two and some only one. Then simple arithmetic shows that the number of people who choose eggs but *not* tomatoes can never be greater than the category of those who choose eggs but not bacon plus those who choose bacon but not tomatoes. Take, for example,

Eggs only: 10 people,
bacon only: 8 people,
tomatoes only: 7 people;
eggs + bacon: 5 people,
eggs + tomatoes: 9 people,
bacon + tomatoes: 12 people;
all three: 4 people.

In that case, 15 people choose eggs but not tomatoes. By contrast, 19 people choose eggs but not bacon, and 13 choose bacon but not tomatoes, making 32 in total in the second category. And 15 is less than 32. Doesn't matter what numbers you pick, this inequality always holds: it's a matter of basic logic. (Try some other numbers for yourself or use a Venn diagram.) It will apply even if the choices are made at random, such as by tossing a coin, because eggs, bacon and tomatoes are real objects that exist before you count them. If photon polarizations were likewise real (but random) ahead of measurement, then they too would be subject to this inequality. For example, if Alice and Bob randomly choose from three crystal orientations at 120° from each other (analogous to the three items of food) they should get coincident results about ⅓ of the time. But quantum mechanics predicts a measurably different result. And experiments confirm the quantum predictions.

The definitive experiment

It took two decades after Bell published his theorem but, eventually, the actual experiment was performed to general satisfaction, using polarized entangled photons. To rule out occult long-range effects, the polarizing crystals and photon detectors needed to be well separated but also to complete their measurements super-fast, so that even at the speed of light there wouldn't be time for any spooky signals to travel between the detectors before the results were recorded. And if the system was nimble enough, then Alice and Bob, being located at opposite ends of the lab, would have time to suddenly change their minds about which polarizer angle to choose, after the photons had already set out.

By 1982, it became possible to achieve these requirements

in the lab. In a series of experiments, Alain Aspect at the École Supérieure d'Optique in Orsay used polarizers positioned 13 metres apart. Measurements were made at various angles. Since Alice and Bob were not available for hire, a high-speed automated redirection system was used for the 'changing their minds at the last moment' part. To cut to the chase: the experiments decisively showed that Bell's inequality was violated. One photon seemingly 'knew things' about the randomly-chosen measurements made on its partner photon in excess of what was logically possible if the photons had all along possessed fixed properties that were completely 'known to each other.' However complete the information about the partner *particle* may be, it wouldn't include what specific *measurement* that particle would be subjected to on the far side of the lab. Yet quantum mechanics predicts, and the Aspect experiment confirmed, that there is indeed such a spooky 'knowledge-at-a-distance' about the measurement made on the distant partner particle.

The original experiment conducted by Aspect and his colleagues was the first of many. Later variants tightened up the experimental protocol and sharpened the results; they also extended the separation of the particles to as much as 30 km. Half a century after Einstein, Podolsky and Rosen threw down their challenge, the definitive experimental test had ruled in Bohr's favour. Both Einstein and Bohr were, by this time, long dead. It took another four decades for the field of quantum non-locality and entanglement to be recognized with a Nobel Prize being awarded to Aspect and others, thirty-two years after Bell's death.

What, then, can we conclude? Most commentators say

that Aspect's experiments demolished the theory of hidden variables, proving that Einstein was wrong and Bohr was right. It is also claimed that we are now forced to choose between faster-than-light physical effects in order to retain a classical version of realism (i.e. that objects have definite physical properties before being measured), or rejecting superluminal influences and thus abandoning classical realism. In my own case, I opt for the latter, which seems to be the majority position among physicists, but there are many nuanced positions. It's not totally clear-cut, and the entire field continues to attract lively debate. Bell himself didn't regard his work as a sort of 'no-go' theorem for hidden variables, but rather as a mathematical criterion that any future theory based on some form of realism would have to satisfy.

How thick is a brick?

Aspect's experiments provided a direct way to discriminate between quantum mechanics and hidden variables, and found in favour of the former. But decades earlier, a handful of theoretical physicists had arrived at a similar conclusion based on pure thought alone. Their arguments hinge on a careful consideration of what physical quantities can be measured together. Not only do they provide a graphic example of the weirdness of quantum physics, they also open the way to some glittering technological possibilities.

To explain the general idea, let me replace atoms – the building blocks of matter – with more familiar building blocks – house bricks. Suppose there is an unknown type of brick concealed in a box, and you can carry out a limited number of specific measurements on the brick to characterize it. For example, you might be able to measure the length, width, height and weight. Well, a brick is a brick: it will already have a definite length, width, height and weight irrespective of which properties you choose to measure, in which order, or at the same time. Imagine, however, that you measured the width of the brick at the same moment as measuring its height and got 50 cm, but if you instead measured the width when also measuring the length you got 60 cm. If the everyday world were like that, it would be a madhouse. Yet a madhouse is what you get in the quantum realm if you insist that particles possess a full complement of properties, brick-like, in advance of measurement. That's because quantum properties, like position and momentum, can be incompatible – they get in each other's way (p. 28). If you measure position first, then momentum, you get different results than doing the reverse – momentum followed by position. The upshot is that the results of quantum measurements depend on the *context* – on what gets measured with which, and when. Consequently, any pre-existing values for things like speed would have to be different depending on which *other* properties of the object you might choose to measure alongside it. They would have to be 'contextual'. Which is pretty profound. 'Contextuality may be the most quantum thing about quantum mechanics,' according to the philosopher Jeremy Butterfield.[7]

The quantum contextuality argument was presented in a mathematical proof by Simon Kochen and Ernst Specker in 1967. Their argument goes like this. Some observables are compatible – for example, energy and momentum – others aren't, like momentum and position. Suppose observable A is compatible with observable B. You can measure them in either order or even perform a joint AB measurement, and get consistent results. Now suppose A is also compatible with C. Same deal – measure either A then C, or the other way about, or AC jointly. Where it gets interesting is the case that A is compatible with both B and C, but B is *incompatible* with C. This can happen, for example, where particle spins are involved (see p. 138). Under those circumstances, it may simply be impossible to replicate the results of quantum mechanics with a 'deeper reality' theory, like hidden variables, that assigns definite *fixed* values to properties A, B and C in advance, independently of what else you choose to measure, i.e. the context. Suppose you believe that A already has a particular value when you decide to measure it alongside B, but then you change your mind and measure A with C instead. In that case, you'd have to assume that A had, all along, possessed a *different* value! Either that, or the quantum particle somehow knew you were going to change your mind and obligingly flipped its properties. If that's an attempt to cling onto some form of objective reality, who needs it?

The magic square and pseudo-telepathy

Once you stop worrying about pre-existing reality and learn to love quantum mechanics, absurdities like shape-shifting bricks get resolved. An entertaining way to see this is in terms of a simple game – reminiscent of the parlour game noughts and crosses. This time the game is played by two people, Alice and Bob, our indefatigable assistants, who collaborate; they either both win or both lose (as in Contract Bridge). Here is what they have to do. There is a square divided into nine blank cells: see Fig. 11(a). The players can write either +1 or −1 in the cells as

(a)

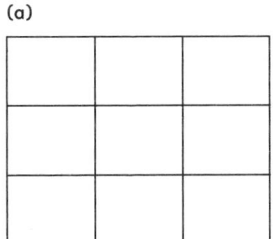

(b)

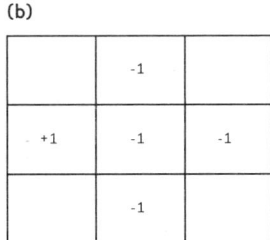

Figure 11

Pseudo-telepathy with the magic square. The square is a grid of 9 cells (a). Alice and Bob play a game by writing either +1 or −1 in the squares, Alice across one row and Bob down one column. Which row and which column are assigned randomly by the referee, Charlie, and neither Alice nor Bob knows the other's choice. The pair win a game if Alice's row contains an even number of minuses and Bob's column contains an odd number of minuses, but only if they write the *same* number in the intersecting square. For example, the pair would win if they chose the assignments shown in (b). But it is impossible to fill the entire square consistently according to the rules. This logical mismatch implies that on average 1 out of 9 games will be lost. If Alice and Bob can use quantum entanglement, however, they could achieve 100 per cent success by making their numerical entries according to the outcomes (+1 or −1) of their respective measurements of a carefully crafted entangled quantum state.

they see fit. The rules of the game state that Alice fills in only rows and Bob fills in only columns, such that there must be an even number of minuses for Alice and an odd number for Bob. Alice and Bob have to play 'blind' – let's say they are put in separate rooms, and cannot communicate once the game starts. To begin, a referee, Charlie, picks a random row for Alice and a random column for Bob. (The choice of row and column constitutes 'the context' in the sense of the previous section.) Alice doesn't know which is Bob's column and Bob doesn't know which is Alice's row. Because every row intersects every column somewhere, there will be a common cell for both. Although they don't know which it is, they need to pick the same symbol, +1 or −1, for the common cell or they lose that round of the game. If they get lucky and pick the same symbol, they win the round if there is one row and one column completed correctly (as in Fig. 11(b)). In the absence of telepathy, it is of course a guessing game, and the pair might get a lucky streak, but statistically they can't win every game, for a simple reason. It's impossible to fill in all nine cells consistently while complying with the rules; whatever patterns the pair go for, there is always one square where Alice and Bob would have to enter contradictory numbers. (Try completing all nine cells for yourself according to the rules and you will inevitably encounter a problematic cell.) There's a 1/9 chance that Charlie will randomly choose that problematic cell for the row-column intersection, so that round would be lost. Thus, in the absence of communication during the game, Alice and Bob cannot beat 8/9 odds overall. The game is known as the Peres-Mermin magic square after its originators.[8]

Now for the magic. In a quantum variant of the game, Alice

and Bob can *always* win. This is how. Suppose that, before the game starts, the players share a particular entangled quantum state, provided by referee Charlie. Instead of choosing for themselves what to write in their assigned squares, Alice and Bob base their entries on the results of certain prescribed measurements on the shared entangled state, measurements that are designed to yield either +1 or −1. Although they won't know in advance of performing the measurements what those numbers will be (because of quantum uncertainty), they are reassured by the entanglement that there will be 100 per cent consistency between their respective entries if they faithfully transcribe the measured results into the cells. Thus, with an appropriately specified entangled state and suite of measurements, Alice and Bob can win the game every time. Entanglement enables Alice and Bob to exploit the quantum world's *lack* of concrete reality to win the game with 100 per cent success. And if they were gambling for money against odds based on the assumption of a pre-existing real world (which is doubtless the belief of most casino managers), they would break the bank. In 2022, a group in China 'played the Magic Square game' (for the good of science, not for money) using quantum computers and verified that the strategy does indeed actually work.[9] The ability of Alice and Bob to cooperate using quantum entanglement and unfailingly win the game without actually communicating has been dubbed 'pseudo-telepathy'.[10]

The genesis of Quantum 2.0

The reality wars fought in the 1920s and 30s by quantum physicists against a dwindling band of sceptics led by Einstein rightly take their place in history as one of the great intellectual tussles of the scientific era. They may have lain for ever as a philosophical footnote to physical theory had it not been for the steady technological advances that Quantum 1.0 spawned: better lasers, faster detectors, improved clocks. That technology had within it the seeds of a new quantum revolution, enabling as it did the landmark Aspect experiment. By the 1980s, entanglement, once regarded as merely a contrived state invoked for argumentation only, began to seem like a precious quantum resource, alongside superposition and contextuality. These quintessentially quantum properties – which have no classical counterparts – opened the way to the creation of previously unimaginable technology and devices with breathtaking capabilities.

By the end of the twentieth century, the outlines of a new quantum revolution were clear, and governments and businesses began to take an interest. The world was already dominated by information technology. But the information being processed in computers, the internet, AI and mobile phones was – and is – strictly classical (binary digits, or bits), even if the devices that manipulate the bits are based on quantum principles. However, quantum concepts like entanglement and superposition transcend classical bits; quantum states embed information in very different and much more powerful ways. The prospect of *quantum information processing* opened up a window into an exponentially greater domain of possibilities, a new reality space of seemingly limitless potential. As we pass the centenary of quantum mechanics, Quantum 2.0 has become a scientific and industrial juggernaut.

Box 3

A bluffer's guide to quantum physics

Here are the essentials:

- **Uncertainty.** Measurable physical quantities, termed 'observables', do not possess well-defined values until measurement. Some observables, called compatible, can be simultaneously measured and have sharp values together. Others (incompatible) cannot; the measurement of one disturbs the measurement of another. In most cases, the uncertainty can be thought of as the relevant observable undergoing random fluctuations in value.

- **Complementarity.** A fundamental feature of quantum mechanics is that it describes how certain physical properties of a quantum system, such as position and momentum, or particle-like and wave-like behaviours, cannot be manifested at the same time or fully known with arbitrary precision. The concept of complementarity was introduced by Niels Bohr as part of his so-called Copenhagen interpretation and highlights the dual aspects of quantum entities, which can exhibit different, seemingly contradictory properties depending on the type of measurement performed.

- **Determinism/indeterminism.** A deterministic system has the property that if everything is known about it at one time then everything at a later time is totally determined. Thus, complete input data can be used (in principle) to correctly predict the state of the system at a later time. Quantum mechanics is indeterministic because complete knowledge of the initial state of a system is usually consistent with a range of different possible measurement outcomes at a later time.

- **Wave function.** This is an abstract mathematical object (function) that completely describes the state of a quantum system. Although it changes with time in a deterministic manner in accordance with Schrödinger's equation (or its generalizations) it does not in general predict the outcomes of specific individual measurements.

- **Born rule.** When a measurement is made, the Born rule offers a simple mathematical procedure to work out the relative probabilities of the various possible measurement outcomes for a given wave function. At the point of measurement, the wave function changes abruptly (it is said to 'collapse') to reflect the suddenly acquired knowledge of the measured value.

- **Superposition.** Many different wave functions can be combined together, reflecting the fact that multiple values of observables can co-exist as 'potential' realities prior to measurement.

- **Entanglement.** Wave functions may describe more than one particle together. One type of joint wave function, called entangled, describes a state in which two (or more) spatially separated particles nevertheless retain a link, which Einstein called 'spooky'. Entanglement implies that measurements made on one particle, even though its properties prior to measurement are uncertain, is weirdly correlated with independent measurements made at the same moment on the other particle.

- **The measurement problem.** The quantum world is characterized by the foregoing properties, but the everyday ('classical') world is not. A quantum measurement interrogates the quantum world but delivers a result in classical terms. An unresolved challenge is to understand how the quantum and classical worlds fit together, given that the classical world is made up of quantum systems.
- **Planck's constant.** Introduced by Max Planck as part of the original quantum idea, Planck's constant is a number that expresses the scale of quantum effects. In familiar everyday units of energy and time, that number is extremely small, indicating that quantum effects are mostly confined to the microworld.

PART TWO

New Technological Marvels

CHAPTER 5

Quantum Information Magic

When I was a student, using a computer entailed trudging across London through the rain with a pocket full of punched cards. The computer itself was housed in its own building, several streets away from my college. Having spent days immersed in complex mathematics scribbled in notebooks, I needed the monstrous machine to help solve the equations for my PhD project. After delivering the program by hand in physical form (in those days there were no terminals to type in the commands), I then had to return a day later to pick up the output, printed in code on huge reams of paper, more often than not betraying a programming error, thus necessitating a repeat of the whole tedious process. Today, the compact laptop on which I am writing this book is far more powerful than the 1960s machine that filled a whole room. The dizzying increase in computational power, and the associated decline in cost, was foretold by Gordon Moore, the CEO of the company Intel, in 1965. Moore predicted that, roughly speaking, the number of components squeezed onto a chip would double every year or two, so that the performance of computers would grow exponentially. Moore's prediction has held fairly steady over several decades and is often dignified by referring to it as 'Moore's Law'. It isn't a law in the physics sense, more a product of innovation and market forces, but as a quantifiable trend it has stood up remarkably well.

The astonishing progress in computing and information technology in the past half century is due in large part to quantum mechanics, which gave us the transistor, then the integrated circuit, then the microchip. The connectivity of the internet is made possible thanks to laser photons transmitted along optical fibres. All this is familiar, of course; it is part of the modern world. But now there is a further transformative leap: the application of quantum mechanics to the concept of information itself. Arthur C. Clarke, the science fiction writer, once wrote that a sufficiently advanced technology would be indistinguishable from magic.[1] The power and scope of the quantum technology now emerging is so dazzling it does indeed seem like magic, even to a professional physicist such as I: a handful of atoms that can outsmart a supercomputer, telepathic data transmission, totally unbreakable codes, sensors that can measure the magnetism of a single atom. All these mind-boggling applications of quantum theory are, however, merely a prelude to what lies on the horizon.

From bits to qubits

To understand the quantum information revolution now surging around us, it helps to start with a very basic question: what, exactly, *is* information? We use the word informally all the time, but to get to grips with the concept of quantum information it's necessary to have a precise definition. The classical one is simple enough: information is reduction in uncertainty. If you flip a fair coin, it may come down heads or tails. Without looking, the result is uncertain, though the odds are known to be 50-50. When you look, the two possibilities, heads or tails, collapse into a single outcome: the uncertainty is removed. The unit of information thus gained by inspecting the coin is called the binary digit, or bit, for short. What's important to grasp is that, although you don't know whether the coin has landed heads-up or tails-up until you actually check, it is surely already showing *either* heads or tails before you inspect it.

When it comes to quantum mechanics, the situation changes completely, because in the absence of a measurement the coin can be in a superposition of *both* heads and tails together. (Maybe not a real coin, but an atomic counterpart.) To analyse how this makes a difference, the classical 'bit' needs to be replaced by its quantum equivalent, known as the qubit. A qubit is not itself a quantum particle; rather, it is the quantum *state* of a particle, or more generally of a quantum system. For example, a qubit might be a photon in a superposition of horizontal and vertical polarization (see Box 1). That qubit could then be manipulated by flipping or rotating the polarization directions around a certain axis using standard optical devices, and this can be designed to enact basic logical operations, the starting point of all information processing. Although the definition of a qubit is technical (and involves complex numbers), a rough analogy is with position on the surface of the Earth. Imagine an explorer roaming the planet, whose location remains uncertain unless he pops up at either the North Pole or the South Pole, with equal likelihood. The question 'which pole, north or south?' is like the heads or tails of the flipped coin, so represents one bit of information. By contrast, a qubit would describe the location of the explorer at any point on the

Earth's surface. It therefore has intrinsically greater scope for encoding information. It's hard to get too excited about a single qubit – which serves as a sort of atom of quantum information. The true power of quantum information processing comes from the possibility of entangling many qubits, which has exponentially greater information content than just combining classical bits. It is entanglement – 'spookiness', in Einstein's parlance – that opens up the vast universe of quantum possibilities.

At first sight, it might seem that quantum effects would be an impediment to information technology. After all, the essence of IT is accurate information management. Yet a hallmark of quantum mechanics is uncertainty. Uncertainty implies errors. To be reliable, the output of a computation needs to be dependable. Remember, however, that uncertainty arises in quantum mechanics only at the measurement step (see p. 27). Left completely undisturbed, the quantum *state* – the wave function – evolves predictably and deterministically, in accordance with Schrödinger's equation. Therefore, if the requisite process, for example a computation, is encoded in the quantum dynamics itself, and remains undisturbed to work its way through, then the result that is read out at the end should be free of errors. In theory. However, achieving error-free fault-tolerant quantum information processing is currently proving a formidable challenge. The reason concerns an inescapable vulnerability of all quantum systems: decoherence.

Don't touch anything!

We've seen how making a measurement on a quantum system irreversibly messes it up. But quantum states are so fragile that *any* external disturbance, not just a meddling measurement, is enough to disrupt them. The extreme sensitivity of quantum states to any form of disturbance is the elephant in the room of quantum information technology. All the techno-marvels dreamed up by physicists and engineers are pie in the sky unless the sources of disturbance can be screened out somehow. That usually means isolating the object of interest – an atom, say – in a vacuum, and shrouding the business part of the experiment or device with metal to prevent stray electromagnetic disturbances. It often entails cooling the entire system close to absolute zero, because thermal fluctuations create a noisy environment that can upset the delicate structure of the quantum state. In particular, where the application is exploiting wave interference, it is crucial that the phases of the waves are carefully preserved. Any interaction between the particle and the environment will scramble the phases and wash out the carefully arranged quantum effects. A state with pristine wave phases is known as *coherent*, and the external degradation of it is called *decoherence*. Pretty much all quantum technology research and development involves a never-ending battle against decoherence.

In that respect, photons fare better than charged particles like electrons, because they interact only comparatively weakly with their surroundings. Laser beams, which consist of many phase-locked photons, are the best-known example of a coherent quantum state, and they can travel long distances even through air without too much decoherence. For more precise applications, such as those involving single photons, transit through optical fibres is better. Each year, more and more progress is being made to evade the degrading effects of decoherence. The longer it can be held at bay, or corrected for, the more powerful the technology.

What, then, can that technology accomplish? The range of possibilities is vast, but let me focus on a few of the more noteworthy examples.

Teleportation

In the science fiction series *Star Trek*, Captain Kirk sometimes wandered about on an alien planet. When he was done, he would say 'Beam me up, Scotty!' or words to that effect, whereupon he would miraculously dematerialize and immediately rematerialize inside the SS *Enterprise*. Teleportation – going instantly from one place to another without passing through the space in between – has long been a dream of science fiction writers. But can it really be done?

Well, yes, after a fashion. In 2004, Anton Zeilinger at the University of Vienna succeeded in teleporting a photon across the river Danube (actually, it went underneath the river).[2] More precisely, what Zeilinger teleported was the quantum *state* of a photon – that is, a qubit – rather than the photon as such.[*] Since those pioneering experiments, quantum teleportation has become the great hope for the future of telecommunication networks, and has been applied to electrons and atoms too. The distance has been extended to as much as 1,400 km from a lab in China up to an orbiting satellite. Zeilinger's spectacular feat contributed to his sharing the 2022 Nobel Prize in Physics.

---

[*] Zeilinger was, however, not the first to demonstrate quantum teleportation. That was done by a group in Rome on 4 July 1997, based on a proposal by Sandu Popescu.

Quantum teleportation is accomplished by exploiting the 'telepathic' properties of entangled states, as spotted in 1993 by Charles Bennett and collaborators.[3] Their basic idea is actually very simple (see Fig. 12). Suppose Alice wants to send Bob a qubit as part of an information package. Alice may have created this qubit herself, or received it from someone further along the line – it doesn't matter. A third party – Charlie, say – first prepares a separate entangled state of two qubits, and sends one qubit to Alice and the other to Bob. Alice now has two qubits. Here is what she has to do. First, she blends her two qubits in a carefully prescribed manner; then she carries out a specially designed measurement on the blended state (the specifics of which I shall skip). The second step results, as always, in the destruction of that state (the wave function collapses). However, Alice knows her original qubit can be reconstructed in Bob's lab in view of the entanglement. This final step can be accomplished if Bob knows the details of Alice's measurement. Armed with that information, he can transform the entangled qubit that Charlie sent him into an exact replica of the qubit that Alice wanted to transmit, using an appropriate manipulation. To find out what Alice measured and the result she got, Bob has to resort to old-fashioned communication, say by telephone. For that reason, the information transfer cannot exceed the speed of light. (There is a common misconception in popular literature that it can.) In spite of his chat with Alice, Bob won't know what the qubit at play is, only that whatever it is, it is what Alice wanted to send. Indeed, Bob may simply pass that unknown qubit on to someone else. Thus, Alice's quantum information isn't *copied* at Bob's end; that would violate a sacrosanct rule of quantum mechanics

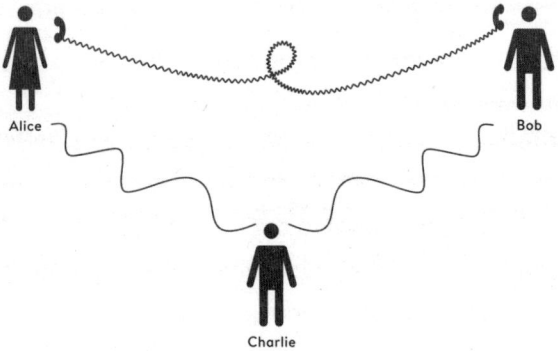

Figure 12

Alice wants to teleport an intact quantum state (a qubit) to Bob. Quantum mechanics forbids copying, so she uses a cut-and-paste strategy instead. First, Charlie sends Alice and Bob one each of an entangled pair of qubits (wavy lines). Alice carries out a special joint measurement on the combined state of her received qubit and the one she wishes to send. Her measurement destroys that state. Bob does not perform a measurement on his qubit or he too would destroy the all-important information. He doesn't know what the information is; his job is merely to reconstruct it faithfully by appropriately manipulating the entangled qubit that Charlie sent him. To that end, he needs Alice to tell him exactly what manipulation to perform on his qubit, based on the results of her (destructive) measurement. Alice calls Bob by telephone to convey those instructions. When all is complete, the quantum information will have been safely transferred from Alice to Bob without passing through the space in between.

called the 'no-cloning theorem', which forbids a quantum state from being exactly replicated. Rather, the state is faithfully reconstructed. In effect, the information transfer is a cut-and-paste job. In spite of the rather convoluted sequence of operations involved in quantum teleportation, it is, as we shall see, crucial to the dream of a quantum internet.

And what of the *Star Trek* scenario? Might a future super-technology permit a human being to be scanned atom-by-atom, then 'cut-and-pasted' – in effect, deleted here and rebuilt there? Would you actually want to risk being deleted? Quantum teleportation implies that your complete bodily wave function, with its trillions of components, first gets massively entangled and then collapsed by measurement, which sounds distinctly unpleasant and would presumably be lethal. The fact that you will be resurrected in another place (if nothing goes wrong) is poor comfort. Somehow, I don't see human teleportation by this method ever becoming a reality.

Quantum cryptography

In real life, information exchanged between private individuals needs to be secure against a malicious hacker or eavesdropper. The time-honoured way of doing this is using cryptography. The message is converted into gobbledegook using a specific code at one end, and decoded at the other using an agreed key. It works well enough unless a snooper manages to crack the code or steal the key. Throughout history, code-makers have been in an arms race with code-breakers. Encrypting information is an indispensable feature of modern life, from banking to spying. Most routine encryption relies on the complexity of certain mathematical manipulations that could make brute-force decryption using supercomputers very time-consuming. But this isn't totally satisfactory. There is a niggling worry that some clever mathematician will come up with a shortcut and fast-track a decrypt; plus, there is always a risk that a hacker will steal the decryption key and unlock the message.

Quantum cryptography, however, opens the way to totally secure messaging because it exploits a *physical* property of quantum systems – the collapse of the wave function. As we've seen, when a quantum measurement is made, the original quantum state abruptly changes by jumping to one of the allowed outcome states, and the information about the original state is irretrievably obliterated. A snooper cannot infer the original state from the measured value and so cannot go backwards and reliably reconstruct that state to cover their tracks. (Remember, measurement is irreversible; you cannot 'unmeasure' something – see p. 36.) So any attempt to eavesdrop by intercepting the information being exchanged is immediately detectable.

This is how one well-known quantum cryptography protocol works. Suppose Alice wants to send Bob some messages in code. She first has to give Bob the key to decode the information. But an eavesdropper (Eve) might intercept the key and be able to read the messages that follow. To circumvent that risk, Alice prepares a series of entangled photon pairs, sending one of each pair to Bob and keeping the other. (They have to avoid the menace of decoherence in doing this.) Alice and Bob then independently make measurements on their respective photons (for example, of the polarization angle, as in the Aspect experiment) using random choices of what to measure each time. The results can simply be expressed as 0s and 1s. Alice then phones Bob and they disclose what their sequence of measurement choices were, but not the actual results (the 0s and 1s obtained) in case Eve is listening in. Because the photon pairs are entangled, whenever Bob has by chance made the same choice of measurement as Alice, he knows the result of his measurement will yield the same result, 0 or 1, as Alice obtained. The pair then discard all the results where they chose different measurement bases, and so end up with an intrinsically random string of 1s and 0s that they know they have in common, which can be used to encrypt their messages by standard digital techniques. If the pesky Eve listens into their conversation, she learns only what the measurement choices were, not the outcomes as 1s and 0s needed to steal the code. But how can the pair be sure that Eve didn't intercept Alice's photons, measure them, and send them on to Bob, who would be none the wiser? Well, any intervention by Eve would wreck the entanglement, and that's easy to detect: for example, Alice and Bob could openly compare the actual results of a random subset of their sequences. If they match, they can be confident that Eve didn't tamper with the process. This somewhat cumbersome method of establishing a totally secure communication channel is in effect 'randomness at a distance' – an example of what is called quantum key distribution.

Quantum cryptography has become something of a boom industry. There are more complicated protocols that include recipient authentication, multi-party communications, error correction, digital fingerprinting and so on. While the field is still largely a theoretical endeavour, there have been some real-world implementations. In 2004, Zeilinger and his team conducted a newsworthy experiment called 'Quantum Cryptography Goes Bankrupt' in which they established a secure communication link between two banks in Vienna using quantum encryption.[4] They employed a quantum key distribution protocol which allowed the banks to exchange secret cryptographic keys encoded in photons. More recently, research institutions, governments, and companies have demonstrated the feasibility of this technology. These experiments have included ground-based links as well as satellite systems, such as the Chinese Quantum Science Satellite (QUESS) named Micius. Some companies now offer quantum key distribution systems for specific sensitive data exchange applications.

The flip side of quantum encryption is, of course, quantum *decryption*. Just as quantum mechanics offers the power to create unbreakable codes, so it has the power to crack conventional codes – a prospect that has sown alarm among governments and a scramble to 'quantum-proof' sensitive communications. The codebreaker's goal hinges on the great hope of quantum technology: the quantum computer.

Quantum computers: the holy grail of quantum information science

In May 1981 the physicist Richard Feynman gave a visionary lecture entitled 'Simulating Physics with Computers' in which he mooted the idea of harnessing quantum processes to enhance the power of computation.[5] Feynman wasn't thinking of codebreaking. What interested him was how to simplify quantum calculations applied to molecules. Schrödinger's equation is readily solved for a simple system such as the hydrogen atom, but when it comes to something even slightly

more complex, say, a water molecule, it is ferociously difficult. It cannot be done by hand (i.e. by a clever student or AI
using standard mathematical functions and manipulations); it
requires a formidable numerical computation on a supercomputer. And the bigger the molecule, the greater the computing
power needed. What Feynman spotted is that a machine using
quantum principles to simulate quantum systems should fare
better. 'Can you do it with a new kind of computer – a quantum computer?' he asked. 'Nature isn't classical, dammit, and
if you want to make a simulation of Nature, you'd better make
it quantum mechanical,' he proclaimed. Feynman did concede,
though, that such a task wouldn't be easy.

It fell to David Deutsch, a physicist at Oxford University,
to write down the formal specifications for a quantum computer a few years later.[6] Deutsch pointed out that such a device
would have a big advantage over a conventional computer, not
only for simulating quantum systems but for many routine
computationally intensive tasks. Deutsch was less interested
in quantum technology as a practical proposition, more as an
abstract concept. The basic principles of a universal computing machine go back to 1937 and a prophetic paper in *Proceedings of the London Mathematical Society* by a young Cambridge
mathematician (and future Second World War hero), Alan
Turing.[7] Titled 'On computable numbers', it addressed a truly
foundational question in mathematics. There are, of course,
lots of numbers – an infinity of them, in fact – and the question
that preoccupied Turing was whether all conceivable numbers
could be 'computed', in the sense of generating the number
after doggedly executing a fixed set of logical operations in
a certain sequence. Turing wasn't concerned with how long

the procedure would take or how many actual steps might be involved, so long as they were finite. An example of a computable number might be the smallest prime number greater than 1,000. If necessary, you could just start at 1,000 and test each whole number in turn to see whether it could be factored. Eventually you would discover that 1,009 is the answer or, to use the modern jargon, 1,009 would be the 'output' of the computation.

Turing had in mind a mechanical contraption that might be adapted to serve as a general-purpose 'computer'. The devices we call computers today are electronic, but the essential point is not what the computer is made of, rather that it is subject to the laws of physics. It can only chug through internal states that are physically permissible. Turing concluded (on purely logical and mathematical grounds) that not all numbers are computable, that is, there exist non-computable numbers, numbers that would *not* be the output of a finite number of pre-assigned fixed operations.

Turing wrote his paper at a time when quantum mechanics was in its infancy, so the machine he envisaged operated according to the laws of classical physics. (Something like a glorified typewriter with a memory tape.) But Deutsch began his scientific career working on problems in the foundations of quantum field theory, so when he began deliberating on the nature of computation, he asked the question: what if Turing's machine operated by the laws of quantum mechanics rather than classical mechanics? Could a quantum computer do things beyond the reach of a classical computer?

It took many years before Feynman and Deutsch's visionary proposal was investigated practically. The basic procedure with performing a computation is to input some data, manipulate it according to a set of rules, and output the answer. Each of these three actions involves information. The set of rules would normally be described as a program, and the whole procedure is now encompassed in the concept of software, to distinguish it from the physical stuff – the hardware – of the actual computer. To make progress with quantum computation means finding a way to embed or encode information in quantum mechanical systems. That step involves replacing the classical 'bit' of information with the quantum 'qubit' that I described on p. 91 The transition from bits to qubits is simple enough to prescribe, yet it opens a whole new world of computational power and unleashes the full potential of the quantum information revolution.

The amazing abilities of a quantum computer stem from the basic properties of superposition and entanglement. To give the flavour of how this makes a (big) difference, consider flipping three coins. Inspecting each coin individually – heads or tails – yields one bit of information per coin, as I explained. Inspecting three coins together yields three bits of information. Alternatively, one might say that three coins together encode three bits of (classical) information. Classical 'coin states' are states in juxtaposition – each coin state is independently heads (H) or tails (T) without reference to the other two. In quantum mechanics, coin states (H or T) can be amalgamated in many different ways. The 8 different combinations $HHH, HHT, HTT, HTH, THT, THH, TTH, TTT$ can be combined in a superposition of any admixture desired, for example, a state could have an equal (1/8) mix of all eight combinations. Now $8 = 2^3$, which means there is exponentially more content in the quantum superposition of three coins than in the classical juxtaposition, i.e. 8 versus 3 in this example. For four coins you get 16 versus 4 because $16 = 2^4$. And so on, with escalating advantage as more qubits are added. (Of course, I'm not referring to actual coins, but their quantum analogues, such as spins – see Box 5.)

Practical challenges

Merely spotting that quantum states can exponentially improve information processing is only the first step. Some major practical hurdles have to be crossed before the dream of a quantum computer can be turned into a functioning machine that can go on the market. The first hurdle is to determine the right hardware to use for encoding the qubits. There are many competing designs out there. One uses single ions trapped by carefully crafted electric and magnetic fields, and then employs the energy levels of the ion as the qubit states. Another is to flip atomic nuclei using their magnetic fields, like minuscule switches. There are also methods involving superconducting loops or nanoscale crystals known as quantum dots that trap electrons in discrete energy levels (see p. 136). Whatever is used, a single qubit is no real help – many of them have to be linked together via quantum entanglement to perform a computation, which involves a sequence of delicate steps. And while in principle that's straightforward, in practice it is a huge struggle against the ever-present menace of decoherence. It's not enough to screen out external disturbances while the information of interest is being fed in, the system has to remain relatively undisturbed for the entire duration of the computation. That proves to be extremely difficult, and the whole enterprise would be a lost cause were it not for the fact that the inevitable errors that creep in as the system evolves can, in some circumstances, be corrected. Elaborate error correction protocols are a big part of getting quantum computers to function as planned. There is also the problem that when all the coherent steps in the computation have been performed, the answer has to be read out, and that step will collapse the wave function and introduce some randomness in the answer. Fortunately, there are workarounds; for example, it's easy to confirm whether a number has been correctly factored into primes by multiplying the output numbers to check whether it yields the given input. There are also statistical tricks to build confidence in the output, and randomness minimization by choosing carefully what quantum variable to measure at the output stage. But even when the hardware problems have been ironed out, there is the software issue. To perform a computation, the operator needs to use a customized algorithm adapted to solving the specific problem of interest. Writing algorithms for quantum computers isn't merely a modification of conventional software, but a whole new enterprise in its own right. All these difficulties add up to a major set of obstacles to be overcome before quantum computing can take off in earnest, but it seems likely that AI will soon dominate writing quantum software as it has for conventional computing.

How well are the engineers doing? The quantum computing industry is in its infancy, and as in all competitive businesses, there is claim and counter-claim, and a lot of hype.[8] Success can be measured in a number of ways. One of these is by quoting how many qubits have been entangled. For a long while, it was limited to just a handful. Now the numbers are greater. But a better measure of success is really how well the device does the job overall. Manufacturers refer to 'quantum advantage' to mean the quantum computer outperforms a conventional supercomputer – getting the same answer quicker, in effect. Another concept is 'quantum supremacy', which refers to the point at which a quantum computer can crack a computational problem that is beyond the reach of standard supercomputers to solve within a reasonable timeframe. In 2019, Google claimed to have achieved quantum supremacy by successfully performing a specific calculation using its quantum processor, Sycamore. They demonstrated that their machine solved a mathematical problem in 200 seconds that would have taken a state-of-the-art classical supercomputer thousands of years.

In spite of this early success, there are still significant issues to overcome, such as improving the reliability, scalability and error-correction of quantum systems, before they can be used for most practical applications. There is also the problem of actually programming them for a range of specific purposes. Nevertheless, the design of quantum computers is a rapidly developing field attracting vast resources. In their 2022 Insight report, 'State of Quantum Computing: Building a Quantum Economy', the World Economic Forum (WEF) estimated that over $35 billion per year was going into the effort, a mix of private and public money.[9] The following year, the WEF unveiled a Quantum Economy Blueprint, and estimated $40 billion of public money now flowing to the sector.[10] Nearly 200 start-up companies have emerged dedicated to quantum computing, alongside industry giants like Google and IBM, with the goal of building faster, better, cheaper machines.

The projected payoff is immense. The WEF reports that quantum computing has the potential to tackle not just physics problems, but 'climate change, hunger and disease'. How is that? Well, many of the practical problems facing the world are exacerbated by their sheer complexity. Take climate change. Although the world is measurably heating up, predicting the specific consequences for a given region is notoriously hard. Will the drought-stricken Southwest of the United States get wetter or drier? Will the Gulf Stream become unstable and, if so, when? Will the Asian monsoon rains shift to higher latitudes? These sorts of questions tax the abilities of the best conventional computers, even with the application of AI. But a quantum computer is likely to yield greatly improved results much faster. Or take disease. Quantum computers will be able to analyse the reams of genetic data pouring out of sequencing labs and design personalized treatments based on an individual's genetic profile. And, in the spirit of Feynman's original vision, quantum computers should be able to greatly speed up drug design, or to search for more effective chemical catalysts. Simulating chemistry accurately is much cheaper and quicker than doing vast numbers of real experiments, and if a candidate molecule is thus discovered its properties can then be checked in the laboratory.[11]

Quantum codebreaking

I've explained how quantum mechanics on the one hand enables absolutely secure data encryption, but on the other, a quantum computer can rapidly crack many of the non-quantum codes currently used for many forms of secure communication. It is the codebreaking potential of quantum computers that, more than anything else, has attracted the attention of governments worldwide. The issue here is this: many existing encryption protocols are based on the difficulty of factoring large numbers into primes. Multiplying numbers is far easier than factoring them. Most people would be able to do the following sum by hand in less than a minute: 73 x 137. The product of these two prime numbers is 1,001. But if you were given 1,001 and asked to work out which are its two prime factors (73 and 137), it would take much longer; you'd have to work through many tries until you hit the right answer. The process is intrinsically asymmetric: easy to go forward, hard to go backward. What is true for a person is also true for a standard computer. If a number is long enough that even the world's fastest supercomputer couldn't factor it into primes in less than a few years, then a mathematical code based on such numbers is currently secure. But in 1994, mathematician Peter Shor designed an algorithm for a future quantum computer to factor large numbers into primes in the twinkling of an eye, thereby blowing the security on a big chunk of the world's communication network.[12] For example, it might take a classical computer the lifetime of the universe to crack RSA, a popular encryption scheme that allows people to publicly share keys. It has been estimated that a quantum computer could do the same job in eight hours.[13]

The impending threat, often dubbed the 'quantum apocalypse', has left governments frantically taking countermeasures. On 4 May 2022, the White House issued a 'National Security Memorandum on Promoting United States

Leadership in Quantum Computing While Mitigating Risks to Vulnerable Cryptographic Systems'.[14] The memo, signed by President Biden, discussed 'the prospect of a cryptanalytically relevant quantum computer (CRQC)', and called upon US government agencies to 'prepare now to implement post-quantum cryptography (PQC)'.

Wising up to the issue, the White House made the following observations:

(a) Quantum computers hold the potential to drive innovations across the American economy, from fields as diverse as materials science and pharmaceuticals to finance and energy. While the full range of applications of quantum computers is still unknown, it is nevertheless clear that America's continued technological and scientific leadership will depend, at least in part, on the Nation's ability to maintain a competitive advantage in quantum computing and QIS [Quantum Information Science].

(b) Yet alongside its potential benefits, quantum computing also poses significant risks to the economic and national security of the United States. Most notably, a quantum computer of sufficient size and sophistication – also known as a cryptanalytically relevant quantum computer (CRQC) – will be capable of breaking much of the public-key cryptography used on digital systems across the United States and around the world. When it becomes available, a CRQC could jeopardize civilian and military communications, undermine supervisory and control systems for critical infrastructure, and defeat security protocols for most Internet-based financial transactions.

(c) In order to balance the competing opportunities and risks of quantum computers, it is the policy of my Administration: (1) to maintain United States leadership in QIS, through continued investment, partnerships, and a balanced approach to technology promotion and protection; and (2) to mitigate the threat of CRQCs through a timely and equitable transition of the Nation's cryptographic systems to interoperable quantum-resistant cryptography.

'Quantum-resistant cryptography' here means finding ways to encode data that are immune from Shor's algorithm or similar, and to that end the US government sent out a follow-up memo on 18 November 2022, urging federal agencies to 'migrate information systems' to a cybersecurity environment immune from quantum computer attack.[15] Given the pace of progress in quantum computing and the deteriorating international climate, such an attack could come at any time.

The quantum internet

Once quantum computers become widely available, there will be an enormous incentive to network them. After all, the transformative power of today's digital computers is that they can be linked together to transfer information between machines and to distribute tasks across many facilities. Data can be backed up in the cloud. Our home computers are now little more than nodes used to couple to the world wide web of information storage and processing, including AIs. Linking computers together for data transfer represents a huge increase in computing power and the range of operations that can be achieved. Many of the things we take for granted, such as email, web searches, online commerce, cryptocurrency, TV streaming and uploading pictures to the cloud, stem from the ability to interconnect computing devices anywhere in the world.

The full potential of quantum computing can be realized only if it can be networked on a global scale. Some uses would be for completely secure financial transactions and voting systems, access to cloud-based quantum computing resources and multi-party networking to leverage practical problem-solving such as climate modelling. Nor is a quantum internet limited to mathematical and scientific problem-solving; it could even have novel implications for the arts (see Box 4).

But for the quantum internet to become a reality, a serious obstacle first has to be overcome. Coupling quantum computers to form a quantum world-wide web isn't just a matter of replacing the existing infrastructure. To preserve its quantumness, information sent from A to B has to remain 'coherent'. I have explained how fragile quantum coherence can be and the extraordinary efforts that scientists and engineers must take to shield quantum information (qubits) from decoherence. Maintaining coherence while transmitting quantum information through the air or down an optical fibre (as opposed to satellites in space) is highly challenging and to date the record distance is about 100 km. So, what to do?

History sets a precedent. The original internet was not, in fact, the one built in the 1980s. Rather it was the one built in the 1870s. This was the heyday of the telegraph. Based on a network of electrical cables that spanned continents and snaked across the sea bed, it enabled signals to be tapped out by Morse code and transmitted between receiving stations in

Europe, America and Asia. The ability to relay messages, and especially commercial information like stock and commodity prices, to far-flung parts of the globe, transformed the world's economy. Suddenly, news from Australia and New Zealand was received in Britain within hours, rather than weeks.

In a heroic feat of logistics and engineering, a South Australian team built an overland telegraph from Adelaide to Darwin in 1872, a mere four years after the first European crossing of the continent. In Darwin, the wire linked up with an undersea cable to what is now Indonesia, and thence (via Asia and the Middle East) to Cornwall and on to London. However, the electrical pulses tapped out by the operators in Adelaide were subject to inevitable degradation as they coursed along the overhead wires, so that after a couple of hundred kilometres they were barely discernible. The problem was solved by building repeater stations along the line at regular intervals. The best known of these was at Alice Springs, in the sparsely populated red heart of the country, named after Alice Todd, the wife of the South Australian government astronomer who was tasked with completing the telegraph project. When a faint signal arrived at a repeater station, it was taken down by an operator and then re-transmitted to the next station, and so on, all the way down the line. Eventually, automatic repeaters were installed in cables to boost the signal without human intervention.

The same basic principle is being developed for the quantum internet, but now there is a crucial difference. Because the very act of reading quantum information 'collapses' the state and destroys coherence, it is no use employing a person, or even an automatic device, to 'read and repeat'. In fact, the no-cloning theorem (see p. 97) rules out any copying strategy. As a result, the concept of a quantum repeater is, like so much in quantum physics, a somewhat subtle one. The basic idea is to use a procedure known as entanglement swapping. Entanglement usually involves two photons generated at a common source and flying apart. But it turns out that two photons from completely independent sources can become entangled. This occurs if the two photons are themselves components of an entangled pair, and one photon from each pair is combined with the other and the combination is subjected to a special joint quantum measurement that entangles them. The upshot is that the remote partners of each pair also become entangled, in spite of being previously unaware of each other. It's rather like the wife of family A meets the husband of family B, and the two fall in love and pair up. The spousal entanglements husband A-wife A and husband B-wife B get messed up by the coupling of wife A and husband B. It happens all the time, of course. But the quantum twist is that the abandoned spouses – husband A and wife B – thereby also become entangled (or fall in love, to pursue the romantic metaphor) even though they are far apart and have never even met!

Here is how entanglement swapping can be used to extend the range of a coherent quantum link. Suppose Alice wants to send a qubit to Bob, but because of the degenerating effect of decoherence, Bob is too far away. If Alice and Bob each send a photon to the mid-point, using one member of an entangled pair that they independently possess, the entanglement can be swapped so that Alice's photon is now entangled with Bob's, thus doubling the distance over which coherent quantum information can be transmitted. By adding more and more repeaters (swappers, really) the distance can be extended to a global scale – in theory. Stony Brook University in New York is one player, leading a consortium to build a long-distance 10-node quantum network to demonstrate quantum communication and distributed quantum processing.[16]

Quantum AI: merging two revolutionary technologies

These days, computer networking is increasingly used for Artificial Intelligence. The reason AI is so quick and knowledgeable is because it is running its data analysis across a vast web of interconnected machines. The birth of a quantum internet, with quantum computer nodes and a quantum cloud, would pave the way for quantum AI (QAI), a thrilling concept, but one with unknown ramifications.

There is a curious historical backstory here. IBM released their first commercially available computer, the IBM 701, in 1952. That same year, the first product using the then brand-new transistor was marketed – a hearing aid. The transistor radio would soon follow. With hindsight, we can see that 1952 marked the birth of two momentous technologies, the first based on information processing and the second on quantum mechanics. These two technologies would soon merge to trigger the transformation to a world dominated by the internet, social media, digital radio and television, virtual reality, language translation and all the rest.

Shortly before IBM began selling their 701, Alan Turing, the mathematician who co-invented the concept of the universal computer (see p. 101), published a visionary article which asked 'Can Machines Think?' in a 1950 edition of the journal

Mind.[17] The same year, the science fiction writer Isaac Asimov published his famous novel *I, Robot*. Again, looking back, we can identify 1950 as the birth of the idea of what is now called Artificial Intelligence. Beguiling though the notion of thinking machines might have been, it actually took seven decades before AI began to blossom. Just in the last few years have we seen the birth of serious AI. While it would be an exaggeration to say that we now have computing systems that can truly 'think', they can certainly outsmart humans in almost every way, from writing thought-provoking poetry in a second to producing complex computer code. Translation software has become so commonplace it is now taken for granted. However, it was the advent of chatbots, like Open AI's ChatGPT, that really attracted attention. These systems, also known as large language models, can be rapidly trained to write stories, references, obituaries, job applications, computer code and, well, almost anything that humans can produce. Their success is down to the huge increase in information processing speeds, the ready availability of massive data storage facilities and the vast resources of the internet.

A number of prominent individuals, such as the late Stephen Hawking, have rung alarm bells about the dangers posed by AI. Increasingly, there is a worry that we may be approaching a tipping point at which AIs are so powerful they will endanger human society. It is always hard to forecast the downstream implications of any new technology, but what adds a further twist to concerns about AI is the possibility of implementing it on a quantum computer. Quantum computer intelligence – dubbed 'quintelligence' by the physicist Frank Wilczek – adds the awesome power of quantum computation to the surging capabilities of neural nets and massive data-bases. At the moment the field is wide open.

One way to think about a quantum computer is in terms of its components. An abacus has a couple of dozen beads to serve as a simple arithmetic processor. A conventional super-computer might have billions of nanometre-scale gates in its central processor. The effective parts of a quantum com-puter, however, are not tiny gates and circuits, nor even the very atoms of which the quantum computer is made. Rather, it is the number of entanglement links that is the true meas-ure of the system's prowess. And the number of potential links is *exponentially* greater than the number of its compo-nents. In a quantum computer, all those states could be superposed (blended together) at once. Even a few hundred atoms could theoretically possess such a staggeringly huge number of possible entangled states that there would not be enough objects in the universe to specify 'in plain lan-guage', as Bohr would have expressed it, what they are – a prospect with almost unimaginable consequences. If a quin-telligence had free reign to explore that *entire* system (the whole 'Hilbert space' – see footnote on p. 36) it would be a journey into the unknown. Indeed, it could involve visit-ing or creating or experiencing (depending on whether we regard quintelligence as conscious) states of the system that could not, even in principle, be described or experienced by human beings, or any non-quantum beings for that matter. Nor can we place any meaningful bounds on what it might create or what thoughts or insights it may have. Of course, a conventional supercomputer, if it were powerful enough to be conscious, might also have thoughts and experiences that would be utterly alien to humans. But quintelligence would have an exponentially greater mental space to explore, giving it a godlike intellect by human standards.

Box 4

Quantum music

Although we cannot be transported into Hilbert space to directly experience the quantum universe, we can perhaps flirt with a pale shadow of it, not through science, but art. There has long been an affinity between musicians and computer scientists, dating back to Ada Lovelace, the collaborator of Charles Babbage, who invented the concept of the digital computer in the 1840s. Lovelace envisaged Babbage's 'engine' generating novel musical patterns. Now a handful of musicians is seeking inspiration from quantum computers.[18] Because the quantum universe is so utterly alien to the human intellect, it promises a new way of thinking about patterns, harmony, form, novelty and orchestration. For example, two musicians can become effectively entangled by proxy, each coupled via conventional computer interfaces to entangled quantum computers, which evolve their input according to the laws of quantum mechanics, taking the composition in unexpected directions and feeding it back in a sort of human-machine jazz improvisation. The quantum computer can behave like a musical companion with vast and unknown skills and limitless potential for novelty. Music follows certain well-defined mathematical patterns, but also involves elements of randomness and surprise. Quantum computers can do both: they generate their own form of mathematical expression, and deliver random, unpredictable output when a measurement is made and the wave function collapses.[19]

CHAPTER 6

Sensing the Unseen

Quantum information engineering, despite its compelling prospects, is severely hampered by a critical vulnerability embedded in the very foundations of the subject: the extreme fragility of quantum states. Elaborate isolation methods must be deployed to maintain coherence. But that very susceptibility implies exquisite sensitivity to external disturbances. Turning a sin into a virtue, quantum fragility can be exploited to make sensors of breathtaking capabilities. There are sensors that can read your thoughts, detect infinitesimal vibrations in the fabric of space and help in the hunt for dark matter. By employing entanglement, superposition, coherence, error correction and qubit manipulation, next generation quantum sensors are leveraging all the techniques being developed for the quantum information revolution now taking place. But quantum sensing technology has advanced far ahead of quantum computing and cryptography in practical applications, achieving unprecedented levels of performance in a whole range of devices. Apart from engineering practicalities and funding, quantum sensing has no obvious limit to sensitivity and accuracy; indeed, we seem to be on a path to a form of quantum omniscience, with consequences for society that are likely to be far-reaching.

Although quantum computing and cryptography receive the most publicity, it is quantum-sensing technology that is the most established area of Quantum 2.0. It has huge commercial potential, and is already impacting many aspects of modern life, with uses in the defence industry, mining, geology and healthcare,[1] but most excitingly, it enables scientists to take a deeper dive into the fundamental nature of matter and spacetime. Across the world, start-up companies, universities and government labs are racing to develop new and improved quantum sensors for a wide range of applications. In the UK, for example, five new quantum technology hubs were opened in July 2024, including Biomedical Sensing, Position, Navigation and Timing, and Integrated Quantum Networks.[2] In the United States, NIST – the National Institute for Standards and Technology – is responsible for much cutting-edge sensor research. Other countries have similar facilities.

If there is a theme that encompasses all quantum sensor technology, it is accurate timekeeping. Time is, after all, the unifying parameter around which physics is organized, as Galileo and Newton realized centuries ago. And historically, the first quantum-sensing devices to see widespread use were in fact quantum clocks.

The best clock in the universe?

If you've ever watched an old-fashioned celluloid movie film run on a projector, it looks like a seamless flow of events. Yet the film itself is divided into static images displayed at about 25 frames a second: we don't notice the joins. In fact, the finest duration that humans can discern is no better than about 10 milliseconds. Clocks can far outperform us on that score.

When I was a child, quartz crystal watches were touted for their accuracy, to about half a second per day, each 'tick' being a mere 0.03 of a millisecond. But this impressive performance pales by comparison with quantum clocks. The original 'atomic' clock was produced in 1949, and used vibrations of ammonia molecules, which have a very precisely defined frequency. The molecules were coupled to microwaves of matching frequency to produce sustained oscillations that could then be read out. The accuracy of this device was impressive: about one second in 300 years. Today's quantum technology, however, far outperforms the ammonia atomic clock.

There are several alternative designs of quantum clock in current operation or under development, and researchers are vying to make the most accurate clock in the known universe. A leading contender is the optical clock, which uses light instead of microwaves, giving it orders of magnitude greater accuracy over the original atomic clock. At its core, the optical clock features a strontium atom that can make transitions between two very narrowly defined energy levels. A highly stable laser is used to induce these transitions. The driving laser and the atom are then locked in a feedback loop at a very precisely defined frequency that persists with perfect regularity over an astronomical number of wave cycles. Each cycle serves as a tick of the clock, which in the case of strontium is less than one hundredth of a trillionth of a trillionth of a second. That's much too fast to read out tick-by-tick, so further carefully designed and stabilized processes have to be added to check the time, so to speak. Additional accuracy is obtained by trapping the atom with lasers and slowing it to a walking pace to cut down the Doppler shift – the very slight effect of an atom's movement on the absorbed frequency of the laser light. Yet further precision can be attained by arranging many strontium atoms in a lattice. The upshot of all these refinements is a clock that is accurate to less than a second in the age of the universe.[3]

Another quantum clock design is based on the time it takes for atoms to free-fall in the Earth's gravitational field. It's an arrangement known as an atomic fountain. The atoms are first slowed to a crawl using a laser, and then given a precise vertical kick, whereupon they rise to a height of a metre or so before falling back down again. They take a fixed time to go up and down, which can be measured very accurately using laser pulses to record their position. Because the atoms are in free fall for about a second – a relatively long time compared to atomic processes – the precision of the measurements is staggering. It has been estimated that the world's best atomic fountain clock has an error rate as low as one second in 40 billion years.[4] Yet even that is not the last word. Work is proceeding on still more accurate quantum clocks, which work by inducing transitions between energy levels in the nuclei of a rare isotope of thorium. Whereas atomic and optical clocks exploit transitions between electron orbits, a nuclear clock would operate at much greater energies, hence frequencies. Also, by using neutrons rather than charged particles for the transitions, disturbances from any stray background electric and magnetic fields are eliminated. Theoretically, such a timepiece could be accurate to one second in 300 billion years.

What's the point of these superclocks? One important application is to financial markets: quantum clocks are widely used by market traders to ensure accurate time stamps for stock market transactions.[5] In the world of high-speed elec-tronic trading, a microsecond can make the difference between profit and loss. Looking ahead, it seems probable that quan-tum computers will replace conventional computers, not just for market analysis, but for actual trades. Because quantum computers process data exponentially faster than convention-al supercomputers, the quantum internet will require ultra-precision timing to manage the immensely rapid transmission of data. These systems will greatly increase the speed and power of informed decision-making and optimization, enab-ling on-the-fly, split-second portfolio adjustments. No wonder the financial services industry is recruiting physicists like crazy.

But quantum clocks have many uses in engineering too. For example, they form the core of the most advanced navi-gation systems. Timing how long a light or radio signal takes to travel between known reference points is a straightforward way to pinpoint one's location, and is the basis for the Global Positioning System (GPS), which combines timing data for signals sent to a network of orbiting satellites, each of which contains quantum clocks. Current GPS enables car drivers, aircraft and drones to navigate to within less than a metre. We take GPS for granted, but there is increasing unease that the system is vulnerable to jamming, spoofing or satellite de-struction by a hostile player, so backup systems are urgently needed. Another problem with GPS is that it doesn't work underground or underwater. It offers little help, for example, in drilling long tunnels or for submarines navigating the depths. What is needed is a transportable device that always knows where it is located without the need to couple to an ex-ternal network. Instruments of that nature are called inertial navigation systems because they work by detecting local movements, not by exchanging signals.

Gyroscopes are a well-known example of a portable navigational aid, and once again, quantum mechanics can enormously enhance their performance. One design is based on an effect discovered as long ago as 1913 by the French physicist Georges Sagnac. If beams of light are sent around a loop in opposite directions, they can be combined to detect if the system is rotating. The same principle is employed in the ring laser gyroscope and the fibre optic gyroscope, commonly used for navigation in aircraft, submarines, spacecraft, robots and autonomous vehicles. They have also found applications in drilling precision boreholes, and in seismology to detect very slight twisting movements of the ground during earthquakes.

A different type of inertial navigation system is the accelerometer, sensitive to any changes in motion. Combined with quantum clocks, quantum accelerometers can attain phenomenal accuracy. One design uses a laser-cooled cloud of slow-motion rubidium atoms, which is ultra-sensitive to external disturbances: the slightest displacements can be detected by the lurching atoms and measured with another laser. Suppose you start out from a known position, and then log and carefully measure the time of every twist and turn, you can reconstruct your trajectory and deduce where you are moment by moment. It's been estimated that a submarine could pinpoint its location with a quantum accelerometer to within a metre after a day's cruising beneath the waves, without reference to any external feature or directional beacon, making the system proof against jamming.

Clocks have, of course, long been the basis for accurate

navigation. Comparing the times of astronomical events at sea with their predicted times at Greenwich can be used to work out a ship's longitude. In 1714, Queen Anne announced a substantial cash prize for a clock that could determine a ship's longitude to within half a degree on the far side of the Atlantic. The prize was claimed by John Harrison in the 1760s for a carefully constructed marine chronometer.[6] Since then, mariners have progressed from navigating by the stars with sextants and chronometers to using directional radio beams, then GPS satellites, and now with on-board quantum clocks and motion sensors.

Navigation entails knowing where you are. The flip side of navigation is tracking: knowing where someone else is. And there too, quantum-sensing technology is disrupting the field, with enormous strategic significance.

Quantum radar

In 2016, the Chinese government claimed they had developed a 'quantum radar' capable of detecting stealth bombers at a distance of 100 km. Although the report was largely dismissed as propaganda, the concept is based on the credible idea of 'quantum illumination', using entangled photons. It works like this. A source creates entangled pairs of photons. For each pair, one is aimed at the target and the other, called the idler, is kept back and stored. When the outbound photon bounces back a split second later, it is compared with the idler to confirm its identity, and a measurement is made, giving information about the target.

Quantum radar would have a big advantage over conventional radar, where the reflected signal is often very weak and has to be discerned amid background clutter. Worse still, military radars are, like G P S, subject to jamming and spoofing. Quantum radar evades enemy countermeasures by effectively fingerprinting the outgoing photons, so that the friendly reflected photons can be distinguished from hostile ones.

Although easy in principle, quantum radar is plagued by decoherence: a photon propagating through the air and bouncing off a target is a messy business that destroys the delicate initial entanglement. However, that disruption doesn't put paid to the whole enterprise, because there's still enough sibling resemblance in the corrupted reflected photon. Imagine twin brothers: one stays at home and the other goes off on a journey and changes his appearance a bit – growing a moustache and dying his hair, say – before sneaking back home, hoping to be disguised and anonymous. However, there may still be a close enough resemblance for the returning twin to be picked out of an approaching throng by facial recognition software. Same with quantum radar. Experiments show that the reflected photon, although no longer entangled with the idler, is still correlated enough with it for a comparison to be made and its identity to be confirmed – for its face to be spotted in the crowd, so to speak. So even when entanglement is broken, quantum correlations can still serve a practical purpose and improve sensitivity.

The biggest obstacle to producing a usable device is that radar works with microwaves, not optical photons, making entanglement technically challenging. The optical equivalent of radar, called lidar, is easier to work with at the quantum level – and is used, for example, in landscape surveying, monitoring atmospheric pollution and in some driverless vehicles. Quantum illumination would greatly increase the power of this optical technology, but on the downside, lidar suffers from the drawback of having a much shorter range than radar nor can it penetrate clouds. It's too soon to say whether easy solutions will be found for these various technical problems, but if quantum radar or a hybrid radar/lidar system can be made to outperform classical radar, it will have many civilian and military applications.

Perhaps the most exciting use of quantum sensing is to fundamental science itself: elucidating the nature of space, time and matter, and the forces that shape the universe. If the job of sensors is to detect the unseen, then the leading candidate for the most elusive unseen entity goes by the name of dark matter.

Sensing dark matter

In astronomy, the dictum 'what you see is what you get' is very wide of the mark. As long ago as 1933, the Swiss American astronomer Frank Zwicky discovered some galaxies in a cluster that were moving much too fast to remain gravitationally bound if the glowing stars and gas constituted all the matter present. But it took half a century before cosmologists came to accept that some unidentified form of unseen material not only pervades the universe, but actually represents the lion's share of all the matter there is, outweighing normal matter – the stuff that you, me and the stars are made of – by a factor of at least five. Naturally, scientists want to know what it is. Until they do, it is simply referred to as 'dark matter'. The prevailing view is that it is made up of some form of exotic particles coughed out of the big bang that gave birth to the universe 13.8 billion years ago. But direct laboratory searches for such a particle have drawn a blank, and the puzzle of dark matter remains one of the great outstanding mysteries of science.

Although dark matter manifests its presence collectively through its gravitational effects far out in the universe, it should be everywhere; there will be some passing through the solar system, and indeed, through the reader, right now. It is creepy to think that, as you read these words, your body is being penetrated by trillions of these enigmatic primordial particles. But of course, you don't feel a thing; nor are you likely to suffer any ill-effects. The very properties that make this matter 'dark' are precisely its lack of interaction with normal matter. Yet, absent of any interaction, how can physicists pin down what this ephemeral stuff is? One answer could be an indirect approach using quantum sensors.

Dark matter experimenters may have been left scratching their heads, but that hasn't stopped the theorists from proposing several mathematical models. One of these predicts that dark matter will have a very slight effect on the values of the fundamental constants of physics, such as the so-called 'fine structure constant' (a combination of the charge on the electron, Planck's constant and the speed of light to yield a pure number – approximately 1/137). But these same 'constants' (a misnomer since in these theories they are posited to vary in value) also determine the frequencies of quantum clocks, with their vaunted tick-tock regularity. If the fundamental constants undergo very slight oscillations as the dark matter slides by, then the tick-tock periods of the clock will wobble in a distinctive way, with a rate, the theory predicts, depending on the mass of the dark matter particle. By comparing the frequencies of different types of quantum clocks, researchers could detect tiny tell-tale differences that betray the presence of a specific form of dark matter in the vicinity. Some astronomers have predicted the existence of vast walls of dark matter that might sweep through the solar system from time to time. Because dark matter interacts only feebly with ordinary matter, this alarming astronomical event would actually pass unnoticed to human senses. But a network of synchronized quantum clocks would be thrown out of kilter by the gravitational effects of the wall on the passage of time.

Other theories predict that the values of the fundamental quantities might vary systematically over the age of the universe. For example, the fine-structure constant could be gradually weakening with time. Quantum clocks have been used to look for such effects, and have placed a stringent limit on how much change is allowed. Measurements at NIST have shown that the fine-structure constant is not currently changing by more than two parts in ten-billion-billion per year. A great deal is riding on these super-accurate measurements: the slightest variation of this quantity would necessitate nothing less than a wholesale makeover of fundamental physics.

The biggest quantum object in the world

Another theoretical prediction, even older than that of dark matter, was made by Einstein over a century ago, and has also been pursued using quantum technology, in this case successfully, leading to one of the greatest scientific discoveries of our time. One thousand three hundred million years ago, two large

black holes in a faraway galaxy spiralled together, and in a titanic collision, gobbled each other up in a fraction of a second.
The result was a single black hole of sixty-two solar masses,
shaking like a monstrous jelly. The awesome cosmic encounter was so violent that it set space itself quivering, sending out
ripples across the universe at the speed of light. The vibrations
of space are known as gravitational waves, and although they
transport prodigious amounts of energy, their effect on material systems is exceedingly small because gravity is so weak
compared to the other forces of nature. Nevertheless, in 2015,
when these undulations swept across our planet, they were
detected in a world first by a pair of quantum sensors possessing unbelievable levels of sensitivity, capable of measuring
movements less than the width of an atomic nucleus over a
distance of many kilometres: that's a few parts in a trillion trillion. To put it in perspective, it's the equivalent of the thick-
ness of a human hair in the distance to the nearest star.

The system that first detected a burst of gravitational
waves is a pair of giant optical instruments located 3,000 km
apart in the USA. Known as LIGO, for Laser Interferometer
Gravitational Observatory, the detectors are now registering
such events on roughly a monthly basis. Similar systems have
been built in Europe, India and Japan. This is, in simplified
terms, how LIGO works: laser light passes through a beam-
splitter – a type of mirror that divides the light equally into
reflected and transmitted sub-beams. The sub-beams are then
directed down perpendicular vacuum tubes; the instrument
forms an enormous L shape, with 4-km-long arms. At the far
end of each, the light reflects off a large mirror dangling from
silica fibres, back up the tube, where the two returning beams

are recombined to produce an interference pattern (see p. 43). Any passing gravitational waves, which are essentially travelling space warps, will disturb the delicate phase correlations between the sub-beams, because they stretch one arm of the interferometer and shrink the other in an undulatory manner. Their passage shows up as a wiggle in the interference pattern.

To get the interferometer to work at all requires an astonishingly high level of precision in quantum engineering, combined with some clever technical tricks, because the mirror movements sought are actually *less* than the quantum uncertainty of the mirror's position.[7] The fact that it works at all is a tribute to the scientists involved, some of whom have devoted a lifetime of work to the enterprise. LIGO and its sister instruments, along with the Large Hadron Collider, must be considered among the highest achievements of human precision engineering, and they also exemplify the effectiveness of international cooperation extending over decades.

An explicit demonstration of the quantum nature of LIGO was made in 2020 by a team from MIT, who commandeered the system for two hours and turned it into an optomechanical vibrator, causing the laser light and mirror motion to resonate.[8] Photons carry momentum, so when they reflect from a mirror they deliver a little kick. By using a delicate feedback mechanism to adjust the phase, the team was able to use the laser light to damp the oscillations of the mirrors until the system came very close to its theoretically lowest energy quantum state – the ground state. Quantum mechanics is normally associated with systems on an atomic scale, and although in recent years quantum effects have been demonstrated on the somewhat larger molecular scale of nanotechnology (p. 137), the domain is still minuscule by human standards. Yet LIGO is a huge piece of equipment, kilometres in size, and the effective oscillating mass has a weight comparable to that of a human being. Yet, astonishingly, quantum effects are discernible.

Finding hidden treasure

On the subject of gravitation, there's a well-worn story about Galileo dropping objects from the Leaning Tower of Pisa to convince sceptics that all material bodies fall at the same rate, regardless of their mass. Over the centuries, this fundamental property of gravity has been tested in the laboratory to high levels of precision, for example, using carefully manufactured brass and aluminium balls. But these experiments all involve large classical objects. How does the gravitational story play out in the quantum domain? Do atoms or neutrons fall as fast as, say, lead balls?

In 1974, a group of physicists at the University of Michigan decided to find out. They analysed how neutrons accelerate in the Earth's gravitational field, within the confines of their lab. Neutrons behave like waves as well as particles, and the length of the waves depends on their speed. If a neutron falls it gains speed and if it rises it loses speed, which means a neutron propelled upwards to a higher elevation acquires a longer wavelength. This (small) effect formed the basis of the Michigan experiment, which is illustrated schematically in Fig. 13. A beam of neutrons was split in two and then recombined at a higher elevation, to form a matter-wave interferometer. One sub-beam took the high road, the other low road. The experimenters measured the shift in the interference pattern to calculate g, the local acceleration due to

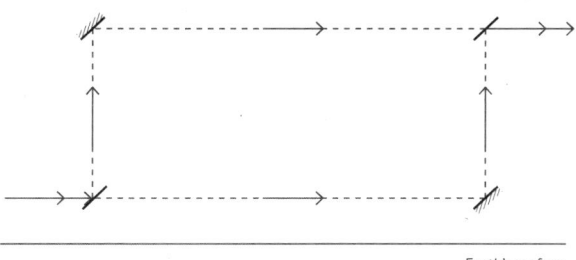

Earth's surface

Figure 13
Quantum measurement of gravity. This highly schematic diagram
shows neutrons entering at bottom left, striking a beam-splitter,
and travelling to a detector at top right along two possible routes –
a 'high road' and a 'low road', in a quantum superposition. Earth's
gravity affects the relative phases of the neutron waves in each
path in a measurable way.

Earth's gravity.[9] And sure enough, they got the same answer as you get from dropping lead balls.

Since the pioneering 1974 experiment, quantum sensors have been used to measure g, the locally experienced force of gravity, to ever greater accuracy. They are now so precise that they can even detect the tiny variations in g that arise from the fact that gravity very gradually weakens with distance from the centre of the Earth, so the pull acting on a particle will be slightly lower the further from the ground it is located. Thus, two entangled particles separated vertically will have very slightly different weights, and engineers can exploit the entanglement to measure the gravity gradient. A French company has marketed a quantum gravity gradiometer using entangled rubidium atoms that can detect changes in g as small as one part in a billion.

Quantum gravity gradiometers have a rosy future in the construction industry. In 2022, scientists at the University of Birmingham were able to locate a tunnel a metre below ground just from the effect of the void on the gravity gradient,[10] and they foresee the use of portable devices to conduct gravitational underground mapping of broken pipes, or the discovery of hidden archaeological sites, or perhaps buried treasure. Quantum gravimetry is also a huge boon to the mining and hydrology industries, for monitoring groundwater changes, seismic disturbances and volcanic activity.

One limitation of 'quantum Galileo' experiments on Earth is that the particles sooner or later hit the ground (or the bottom of the apparatus). But by going into space, much longer free-fall times can be achieved because any particles released in orbit inside a spacecraft are effectively weightless;

they fall around the Earth rather than towards it. In 2018, NASA installed a Cold Atom Laboratory in the International Space Station, and in May 2020, the astronauts were able to demonstrate atom interferometry in orbit using a cloud of ultra-cold atoms.[11] This experiment was more than a bit of quantum fun. According to NASA, it 'heralds a future in which space-based quantum sensors become a widely used tool for scientific exploration of the universe', such as tests of general relativity, spacecraft navigation, and prospecting for subsurface minerals on the moon.[12]

When quantum meets nano

So far, I have described how quantum sensors are being used to probe unseen or previously undetectable effects mostly on an astronomical or human scale. But quantum-sensing technology also has many applications on a microscopic scale too, with great relevance across a range of industries, including materials science, environmental monitoring, solar cell design and medicine.[13] Underpinning this promise is the possibility of fabricating smart molecular devices that can carry out prescribed tasks with high specificity. The basic concept goes back to another seminal lecture delivered by the physicist Richard Feynman at the California Institute of Technology, in 1959. Entitled 'There's Plenty of Room at the Bottom', Feynman considered the possibility of directly manipulating individual atoms and building machines on a molecular (i.e. nanometre) scale. A quarter of a century later, the term 'nanotechnology' was coined for such an endeavour by Norio Taniguchi of Tokyo Science University.

In spite of the far-reaching possibilities, the subject of

nanoscale engineering remained a speculative backwater until the 1980s, when it was propelled into prominence with the publication of Eric Drexler's book, *Engines of Creation: The Coming Era of Nanotechnology*.[14] Drexler's grandiose vision hinged on the possibility of exercising control over matter and forces at the smallest practical scale to, among other things, fabricate novel materials atom by atom, create self-replicating molecular machines, repair damaged tissues and cells, deliver drugs inside the body in a targeted way and clean up the environment. Although many of these ambitious goals have yet to be realized, the field of nanotechnology is now blossoming, with many applications in medicine, materials science and fundamental physics.

When matter is manipulated on a nanoscale, quantum effects will inevitably loom large, which means much of nanotechnology is an application of quantum engineering. An early invention was the quantum dot (see p. 58), a nanometre speck of semiconductor, so small that electrons are confined as if trapped in a tiny box. (The quantum dot was the subject of the 2023 Nobel Prize in Chemistry.) The electrons display discrete energy levels as in atoms and emit light at frequencies that depend on the size of the speck. For this reason, quantum dots are often described as artificial atoms. Unlike real atoms, where scientists have to work with the energy levels provided by Mother Nature, artificial atoms can be customized for commercial purposes to emit light of chosen colours – you see quantum dots at work whenever you look at a high-definition TV screen with its vibrant displays. But they have also found widespread industrial use in applications from medical imaging to solar cells and in single electron transistors.

Quantum dots are effectively zero-dimensional electron traps. But nanotechnology permits electron confinement in one and two dimensions too, producing a wide variety of quantized energy levels, with a host of applications. Quantum wires, for example, are used in nanoscale transistors, integrated circuits and quantum sensors. Carbon nanotubes display a range of novel electrical properties because they confine electrons' movement around their circumference but allow unimpeded flow along the tube. They have many possible applications, ranging from high-strength materials to drug delivery that targets specific organs or tumours. Quantum wells (two-dimensional conducting sheets) are used in semiconductor lasers and infrared detectors. Graphene sheets have a single layer of carbon atoms arranged in a hexagonal lattice, and exhibit a rich variety of electrical, mechanical, and thermal properties with applications in optoelectronics.[15] Nanotechnology can also be used to fabricate devices known as topological insulators, which act as insulators in their bulk but have conducting surface states. These are being used as components in quantum computing. Additional nanoscale structures now available include mechanical resonators – beams, cantilevers, membranes and paddles – that vibrate with quantum energy levels and can be cooled to their ground states.

One compelling area of quantum nanotechnology still in its infancy is the design of engines and heat pumps on a minuscule scale, a field known as quantum thermodynamics. Like their classical counterparts, these devices are made to operate in a cycle, but in place of the working fluids such as petrol vapour or a refrigerant gas like freon, they use quantum states and exploit their novel properties. Sometimes they work with a single quantum particle. Fascinatingly, quantum measurement itself can be used as a type of fuel to power an atomic-scale engine, because the very act of measurement causes a quantum superposition to jump into one of the allowed outcomes, and that jump is often accompanied by a transfer of energy into or out of the measured system. Practical applications include quantum refrigerators, which can be used to cool other nanoscale devices crucial for the operation of quantum computers. But the principal interest in quantum thermodynamics is to test the fundamental limits of physical laws. The Industrial Revolution of the nineteenth century was essentially a dramatic application of the laws of classical thermodynamics at scale. Quantum thermodynamics is not merely a scaled-down version of its classical counterpart, but involves entirely novel principles and possibilities. Combined with quantum information processing, quantum thermodynamics holds the prospect of driving a further global revolution in technology and economics, enabling feats of engineering impossible to attain using classical physics.

Box 5

Spin

Schrödinger's wave equation correctly predicts that atoms have discrete energy levels – certain fixed energies for the orbiting electrons. We say that the energy levels of atoms are 'quantized'. But Schrödinger's equation predicts that other properties of matter are quantized in discrete units too. One of these is rotation – technically, angular momentum. A spinning top, for example, can go at any rotational speed, according to classical physics. But quantum mechanics permits only certain fixed amounts. What came as a surprise is that, in addition to an electron having quantized *orbital* rotation, it also possesses a type of internal rotation, called intrinsic spin. That too is quantized. All particles of matter – electrons, quarks, neutrinos, etc. – have exactly the same quantity of intrinsic spin. Not only is spin quantized, the direction in which the spin axis points is also quantized. Like position and momentum, spin direction is one of those properties that, according to textbook quantum mechanics, remains ambiguous and ill-defined until a measurement is made, when the spin is projected to align or anti-align along the axis of measurement. Whichever angle an experimenter chooses to make the measurement, the particle's spin is always found to point either 'up' or 'down' that axis. Importantly, the components of spin in different spatial directions are incompatible variables, which means they 'get in each other's way' when measured, like position and momentum. Measurements of spin directions play a critical role in much of quantum technology, and they also feature in contextuality experiments and magic square pseudo-telepathy, too (see p. 78).

From electronics to spintronics

Asked to name a property that captures the essence of quantum mechanics at its most basic level, a majority of physicists would pick intrinsic spin (see Box 5). It is the simplest manifestation imaginable of discreteness, incompatibility, uncertainty, and all the rest. Unlike, say, energy, spin measurements can yield only two outcomes: 'pointing up' or 'pointing down', *whichever* direction you choose to measure and designate as 'up and down'. This sets it apart from other forms of rotation, such as orbital motion or the spin of a classical body. In fact, intrinsic spin is so distinctively weird that it requires a whole new branch of mathematics to describe it, a formulation that also links quantum mechanics to the theory of relativity and spacetime structure. Curious readers can visit Westminster Abbey to see the key equation, on a plaque dedicated to the theoretical physicist Paul Dirac, who first put these concepts together in the late 1920s. Because electrons are charged, their intrinsic spin implies they are also tiny magnets that can be manipulated with applied magnetic fields. And flipping between up and down encodes a basic logic operation. These features open the way to the cutting-edge field of technology called 'spintronics' that promises to push Moore's Law (see p. 89) forward for another decade and dramatically slash the energy bills that rampant AI development is unleashing.

The roots of spintronics go back to the 1970s, and a discovery by Michel Jullière, a physicist at the Institut National des Sciences Appliquées in Rennes. Jullière was investigating the tunnel effect (see p. 33) between two electrically conducting magnetic materials separated by a thin insulating layer to

make a sandwich – a routine enough undertaking. As expected, some electrons were able to quantum-mechanically tunnel
through the barrier. What came as a surprise, however, was
that the degree of tunnelling depended on whether the conducting layers forming the 'bread of the sandwich' were magnetized, and in particular whether they were magnetized in
the same direction, or in opposite directions. Armed with the
concept of intrinsic spin, we can see that Jullière's discovery
makes sense. If the 'slice of bread' is magnetized, then that de-
fines a specific direction in space, and, as explained, the spin
of an electron must then point either along that direction or
opposed to it ('up' or 'down'), i.e. either along, or opposed to,
any external magnetic field. As a result, the strength of the
barrier to be tunnelled through depends on which way the
electron's spin is pointing, because that determines whether
the interacting magnetic fields of the layer and the electron
are aligned or opposed. Jullière's humble result immediately
opened up an exciting possibility – of using the spins of electrons, as well as their charges, to manipulate their behaviour.
An early application was for magnetic data storage – the well-
known random-access memory (RAM) in computers. But
with advances in nanotechnology, the subject of 'spintronics'
is now developing rapidly.[16] Because it takes far less effort to
simply flip a spin than to move a particle around, information
can be processed far more cheaply and efficiently using spintronics rather than electronics. Furthermore, the information
stored is more stable and physically compact, and the read/
write speeds are greater.

Spintronics also enables more powerful and efficient
magnetic sensors. And it is the measurement of ultra-weak

magnetic fields that opens the way to one of the most promising but contentious applications of quantum-sensing
technology.

Quantum neuroscience and other medical marvels

The brain is abuzz with electrical activity (it's how we think!),
and the intricate swirling currents generate complex flickering
magnetic fields that reach outside the skull, where they can be
detected by quantum sensors and tracked in real time, enabling
neuroscientists to build up a sort of mental activity map. The
practice goes by the cumbersome name of magnetoencephalog-
raphy, or MEG for short. There are other tools to peer inside
one's head, such as functional MRI (another bit of quantum
technology that can be traced back to 1938), electroencephal-
ography (EEG) and electrode implantation. But MEG is non-
invasive, does not require the powerful applied magnetic fields
of MRI, and has faster time resolution than EEG.

The first MEG recordings were made in 1968, using a
single highly sensitive magnetometer called a SQUID (for
Superconducting Quantum Interference Device). SQUIDs
were the brainchild of Brian Josephson of Cambridge University, who made the critical discovery while doing his PhD. If
a superconductor (see p. 12) forms a ring, an electric current
flowing around the loop produces a magnetic field threading
through the loop. Being a fundamentally quantum system, the
magnetic field strength is itself quantized. If a thin slice of in-
sulator is inserted in the material of the ring, the current isn't
totally shut off because a small amount can still quantum-
mechanically tunnel though the junction (known, fittingly

enough, as a 'Josephson junction'). This provides the ring with exquisite sensitivity to detect extremely small external magnetic fields – as low as a billionth of the strength of Earth's field – by counting the tiny changes in the discrete units of the ring's own magnetic field. The development of multi-channel SQUID systems in the 1980s allowed for whole-head MEG, enabling the simultaneous recording of magnetic signals from different brain regions, making it a powerful tool for studying the highly dynamical electrical patterns.

In recent years, a greatly improved type of quantum magnetic sensor has been developed. Known as an Optically Pumped Magnetometer (OPM) it uses a laser to excite a cloud of rubidium atoms into a magnetically sensitive state. Atomic electrons produce magnetic fields from both their orbital motions and from their intrinsic spin. Nuclei also have magnetic fields. The interplay of these several magnetic effects creates a whole set of closely spaced energy levels. A finely tuned laser can be used to excite the atoms into a specifically selected level, chosen for its magnetic sensitivity. When the laser has energized most of the atoms in an aligned way, the cloud becomes transparent, but the presence of a magnetic field, such as from brainwaves, jumbles the atoms' orientations and makes the cloud murky, an effect that is easily detected with the laser. OPMs incorporated into a wearable helmet can produce magnetic images of the brain's surface areas with millimetre accuracy and millisecond resolution. Manufacturers of OPMs claim that such scanners can help analyse a range of neurological disorders, including autism, epilepsy, dementia and schizophrenia. The University of Birmingham's OPM development project at the Centre for

Human Brain Health foresees combining transcranial magnetic stimulation (TMS) with OPMs, making it possible to excite one part of the brain and measure the response in another part, helping neuroscientists map the internal connections and investigate whether various neurological disorders are associated with changes in the wiring.[17]

Spintronics forms the basis of the most promising type of quantum magnetic sensor: a microscopic diamond containing a vacancy in the crystal lattice, next to which a nitrogen atom is bound.[18] Electrons trapped in this defect structure have quantum states that are exceedingly sensitive to any external magnetic field. There is a small difference in the energy levels of the trapped electron according to whether its spin is aligned with, or opposed to, the external field. By measuring that difference using lasers, magnetic fields can be detected in tiny regions, even on the scale of nanometres. The plan is for these minuscule quantum diamond sensors to produce images of single cells and even of structures within cells, making them promising for medical applications such as targeted drug delivery, early tumour diagnosis and specific biomolecule detection. Quantum diamond sensors can also be used as mini-thermometers to produce heat maps inside living cells, accurate to within a few thousandths of a degree.

Although MEG and other neuroimaging technologies are primarily directed to medical research, there is a growing interest in the subject of quantum assisted transhumanism. If quantum sensors can monitor your brainwaves with high resolution, then maybe you can wear a helmet that can, in effect, read your thoughts. Such a device could be used to couple your neural activity to a computer, giving you the ability to communicate with external systems by thought alone, bypassing traditional pathways like speech or movement. This is the groundbreaking field of Brain Computer Interface (BCI) technology that lies at the intersection of neuroscience, engineering, and computing. By translating neural activity into signals that can control computers, prosthetics, or robots, BCIs have the potential to transform various aspects of human life, from healthcare and rehabilitation to warfare, space exploration and entertainment. Exciting though it may be, BCI technology comes with serious ethical concerns related to privacy, consent and the possibility for misuse, such as hacking brain signals or using BCIs for surveillance. There is also a debate about whether coupling the human mind to a powerful computational system such as an AI robs us of our humanity or enhances our humanity. As quantum-sensing technology and signal processing advance, many difficult questions about regulation and access will need to be confronted.

What next?

This chapter has outlined just a few of the amazing new sensors based on the quantum properties of entanglement, superposition and spin. While I have focused on the weird physics that underlies quantum technology, all the applications I have discussed require the immense skill and intuition of engineers who can turn abstract notions into practical devices. The surge in quantum-sensing applications hinges on a killer combination of the ultra-sensitivity of phenomena like interference, combined with the cheap, high-speed data processing available to us today. This is, of course, only the start. Taking a leaf from Moore's Law, we can imagine that every year or two the pace of progress in quantum tech will double – more entanglement, longer coherence times, cheaper qubits.

Quantum computing and quantum sensing may grow spectacularly in the next few years, but they are just improving existing technological infrastructure. What else might arise that is entirely new? As with all novel and powerful technology, quantum tech will generate its own markets and its own applications. Things we haven't thought of yet may become commonplace in a couple of decades. Expect many eye-opening quantum innovations in the home, the workplace and in space in the near future, as we witness the full reach of Quantum 2.0.

CHAPTER 7

Quantum Biology

The natural world abounds with ingenious mechanisms and systems, and engineers have long sought inspiration from the study of living organisms. Over the years, biomimicry has yielded many applications, from new types of adhesives to underwater drones. Now there is growing evidence that some living organisms harness quantum effects with great effectiveness. This is the emerging field of quantum biology, of obvious interest to scientists and engineers working on quantum technology, but also of profound significance for understanding the nature of life itself. In the previous chapter I described some examples of quantum technology in the service of human biology. Now I want to look at the converse: quantum biology in the service of human technology. This is a subject that can be traced back to none other than Erwin Schrödinger himself.

What is life?

For all its power and scope, the new theory of matter formulated in the 1920s left one subject largely untouched. Within just a few years, quantum mechanics had explained much of physics and chemistry, but biology remained a stubborn mystery – and it still does to this day. What secret sauce gives living organisms their special oomph? How do they accomplish such amazing feats of organization and goal-seeking behaviour?

Two decades after he published his famous equation, Schrödinger took on the challenge, delivering a series of lectures at Trinity College Dublin in Ireland, during the dark days of the Second World War. Entitled *What Is Life?*, the lectures were published in book form in 1944, and became immensely influential in the formative stages of the subject of molecular biology.[1]

Schrödinger spotted that quantum principles would underpin the mechanism of heredity: genetic information must be stably stored and faithfully transmitted at the molecular level. But in many other respects, living organisms appeared to be some form of magic matter. The physicist Max Delbrück expressed it eloquently in 1949:

> The curiosity remains . . . to grasp more clearly how the same matter, which in physics and in chemistry displays orderly and reproducible and relatively simple properties, arranges itself in the most astounding fashions as soon as it is drawn into the orbit of the living organism. The closer one looks at these performances of matter in living organisms the more impressive the show becomes. The meanest living cell becomes a magic puzzle box full of elaborate and changing molecules . . .[2]

Even on casual inspection, it is clear that living organisms have unusual and special properties. And yet living things are made of normal atoms and molecules. Nobody has found any trace of a special 'life force' to explain the magic. It's weird.

But so is quantum mechanics. Could weird explain weird? Is the very essence of life a manifestation of quantum weirdness, or does life transcend even quantum mechanics? Schrödinger was open-minded about the question. One must be prepared to find 'a new kind of physical law' prevailing in living matter, he allowed, but he didn't elaborate.

And there the matter might have rested. For decades, molecular biology advanced apace – the structure of DNA, the cracking of the genetic code, the mechanism of protein assembly – all explained with classical ball-and-stick molecular models. Organisms seemed to be little more than exceedingly elaborate Lego assemblages. Quantum mechanics might be essential to explain the shapes and binding forces of the building blocks, but beyond that, classical physics appeared to suffice. Still, there were a few troubling anomalies, things that didn't quite fit the classical paradigm. By the turn of the twenty-first century, these oddities had accumulated to a point where a handful of scientists began to talk about 'quantum biology' – a sector of the life sciences that demanded the full gamut of quantum weirdness: tunnelling, coherence, superposition, entanglement and spin. While the subject remains controversial, there is good evidence that at least some biological processes harness quantum effects for advantage – and that advantage might be mimicked and engineered into human technology, too.

Photosynthesis: harvesting photons

Photosynthesis is the primary production process that sustains life on Earth, using photons to split water molecules and build biomass. The photon is the quintessential quantum entity, which might seem to make photosynthesis the very definition of quantum biology. But until recently, the photon itself was considered little more than the buck that paid for the biomass bang – just a useful source of energy. One noted property, however, is the extraordinary efficiency of photosynthesis. Very few photons are wasted: almost all of them contribute to the chemical output. Intrigued by how, scientists worked out the complicated steps in the process. Incoming photons are initially collected by specialized light-harvesting molecules. However, the downstream chemical transformations take place in a different part of the cell. The garnered energy needs to be transported as rapidly and efficiently as possible from one to the other, with minimal loss on the way. In the early 2000s, a research group at the University of California in Berkeley, led by Graham Fleming and Gregory Engel, developed ultrafast spectroscopic techniques to probe the inner workings of this energy transfer process. Their results, which came from samples of green sulphur bacteria that use a type of photosynthesis, were surprising. The absorbed photon excites a complex molecular mesh that passes on the energy to the chemical factory used to make the biomass product. The excitations can be described quantum-mechanically as waves, and what Fleming and his colleagues found is that the quantum waves remain coherent, against expectations, within the molecular complex for long enough to do the job. That is, the wave superpositions serve to enhance the speed and efficiency of the energy delivery.

In spite of these exciting results, much more work is needed to convince sceptics that life on Earth is sustained by a primary energy source that exploits quantum weirdness in its critical operation. Moreover, it is unclear how relevant the laboratory investigations are to photosynthesis 'in the wild'. Living things might contain many components that, when isolated and probed under controlled conditions, can be coaxed to exhibit quantum effects that play little or no role in the naturally functioning organism. That doesn't make the effects of zero interest, because even isolated components could form the basis for technological applications. And indeed, biology is the inspiration for a wide range of research on artificial photosynthesis, which aims to replicate the natural process to design better photocells, and produce chemical fuels (such as hydrogen or methanol) directly from sunlight, water and carbon dioxide. By studying photosynthesis, scientists hope to design artificial systems that can achieve similar or even higher efficiencies.

Birds that can find their way in the dark

With the benefit of modern technology, humans are pretty adept at finding their way around the planet. But we are not the only ones able to perform amazing feats of navigation. Ornithologists have long been astounded by the ability of birds to travel across the globe to specific locations. Homing pigeons are a well-known case, but some birds not only seem to know where they are going, they fly enormous distances to get there in a straight shot. For example, a bar-tailed godwit was tracked by satellite flying from Alaska to New Zealand in a 11,000-km nonstop flight across the Pacific Ocean! A directional error of more than a few degrees could have been fatal. How do these birds do it? The answer in many cases seems to be magnetism. Evidently, they have some sort of compass in their heads that can detect the Earth's magnetic field to orient themselves precisely. The puzzle is that the Earth's field is extremely weak – far less than the signal from a mobile phone, for example. There must be a super-sensitive detection mechanism at work, which hints at something quantum. In the last few years, scientists think they have figured out how a quantum magnetic sensor might work in a bird.[3]

At root, the avian compass is a by-product of electron spin. Recall that all electrons possess intrinsic spin; they are like tiny spinning tops (see Box 5). And because they are charged particles, every electron is a magnet. If an atom possesses an even number of electrons, their spins pair up and oppose each other, cancelling out any overall magnetic effect. But some atoms have an odd number of electrons and so they are magnetic. Those magnetic atoms are called 'radicals'. Chemical reactions involving radicals can obviously be influenced by external magnetic fields. For the avian compass, the details are complicated, but the basic idea is that a specialized molecule fixed in the bird's eye at a given orientation gets hit by a photon that creates a pair of radicals, by transferring a single electron from one molecule to another. This energized radical pair might simply revert back to the original molecular arrangement, or it might use its energy to drive a different chemical reaction and make a new molecule. If that product molecule is a neurotransmitter, it could enable the bird to directly sense something.

To summarize the details, there are two chemical exit pathways from the radical pair state, and the ratio of their products, and hence the strength of the signal sent to the bird's brain, depends – so the theory goes – on the direction of the

Earth's magnetic field. This is why. When an electron is subjected to an external magnetic field, its spin direction gyrates (a process called precession). When the radical pair is formed by absorbing a photon, the transferred electron remains entangled with an electron in the donor molecule. Because of the entanglement, the two electrons will gyrate together. But – and this is the critical detail – they will not maintain synchrony because the immediate magnetic environments of the two radicals plus the external field of the Earth will be different. The upshot (if this theory is to be believed) is that the gyrations get out of whack, and that affects the yield of the neurotransmitter at a level that depends on the orientation of the molecules to the Earth's field. All of which makes the bird brain think, 'Ah! Fly that way!'

Obviously, the photo-excitation mechanism won't work in pitch blackness, but extremely low light levels seem to be adequate, such that birds can still find their way at night. The mechanism looks all the more remarkable because the Earth's magnetic field is so feeble: its energy is far below what might directly bring about chemical transformations. It is only the delicate entangled dance between aligned and opposed spin configurations of the electrons in the radical pair that creates a state of ultra-sensitivity to weak magnetism. Curiously, in a strong magnetic field, such as you might find in an MRI machine, the interaction acts so fiercely on both electrons that it completely swamps the finely tuned wobble between the configurations, and obliterates the sensing mechanism. The birds would be 'magnetically blind' on a planet with a much stronger magnetic field.

Although the (necessarily) convoluted explanation above

looks to be a bit of a stretch, researchers have isolated the key molecules and experimented with laser pulses and applied magnetic fields, and the theory does seem to stand up to scrutiny. Furthermore, it turns out that some migrating butterflies also possess similar distinctive molecules. In fact, these specialized molecules are common features of many animals, humans included, hinting that maybe quantum magnetic sensing is an ancient property across the animal kingdom, now mostly defunct or redundant, but honed to perfection in a handful of species.

Given that magnetic fields do affect electron spin states with downstream consequences for chemistry, it would seem that there should be many other examples of organisms affected by, or responding to, magnetic fields. And actually, there are. There is a whole subject of biomagnetism under active study, with subject matter ranging from embryo development, through cancer, to wound-healing. What is unclear at this stage is whether in these other examples there are any fancy quantum mechanical goings-on, or whether the effect of the magnetic field on the organisms is purely classical. Nevertheless, the avian compass could provide opportunity for biomimicry applications, perhaps in the domain of spintronics (p. 140).

Life's quantum internet

A year before Schrödinger formulated his famous equation, a paper appeared in a German biology journal by a Russian embryologist, Alexander Gurwitsch, claiming that dividing cells emit ultra-violet photons. Gurwitsch referred to this as 'mitogenic radiation' (mitosis being the technical term for cell division) and he conjectured that the photons played a bio-logical role by prompting division in neighbouring cells. He observed that mitosis in the tip of one onion root seemed to stimulate division in a nearby root, even when the roots were separated by a glass plate. That organisms emit light was not especially mysterious: fireflies and luminescent bacteria do it all the time. But Gurwitsch's emissions were far more elusive, having very low intensity, typically between one and a thou-sand photons per square centimetre per second. That made the photons hard to detect and study, and many years were to pass before much progress was made on this tantalizing phenomenon.

In the 1970s, the German biophysicist Fritz-Albert Popp, following Gurwitsch's lead, also studied the low-level emis-sion of light by plants, as well as animal tissues and cell cultures, and proposed that these 'biophotons' formed an im-portant regulatory and signalling function between cells. Popp also claimed that the photons carried the distinctive stamp of coherent quantum effects. He arrived at this conclusion by analysing the statistical patterns of cascades of photons. The significance of counting photons over various durations is that those from, say, an incandescent light bulb are general-ly emitted haphazardly, whereas coherent laser light is highly organized. Popp thought he saw the tell-tale signs of quantum organization in his biophotons. It's unclear what the source of the light might be. Some researchers have proposed mito-chondria or sites of oxidative damage. Popp suggested that the photons might come from DNA. All these claims have been challenged and require further investigation.

What might the purpose of the biophotons be? It has been

suggested that stressed organisms emit bursts of photons as a sort of luminescent cry for help. More speculative is that biophoton exchange could form a type of signalling system – a quantum internet for life – to coordinate activities such as gene expression, growth, differentiation and programmed cell death, or to synchronize cellular functions and maintain healthy tissue organization. Signalling between plants, or between bacteria in colonies, could serve as warnings to neighbours to initiate defence mechanisms against common threats. That cells and organisms signal each other is an established part of biology, but it is usually assumed that the messenger agents must be molecules. Photon messaging would have a distinct advantage in being fast and precise, and could in principle form a parallel web of information exchange for regulation and control in, for example, embryo development, limb regeneration and other processes requiring a fine level of management over an extended region.

Quantum brains

There is a long history of speculation that quantum mechanics plays a decisive role in brain function, going back to the Nobel prize-winning neuroscientist John Eccles, who in 1990 suggested that quantum effects occur in synaptic transmission.[4] In more recent years, another Nobel Prize winner, this time the mathematician Roger Penrose, has also postulated that quantum coherence might be operating in neural tissue at the subcellular level, in microtubules. These speculations have been received with much scepticism, but there is one area of neural activity where quantum effects do seem plausible. The way the brain shunts information around involves the transport of ions, especially potassium, sodium and calcium, across membranes. It is remarkable that brains can discriminate between a potassium ion and a sodium ion, given their chemical similarity; they can be distinguished only by a slight difference in their physical size. These ions cross the membranes of axons – the signal-conveying 'wires' linking neurones – by passing through little holes that form narrow channels. The tiny apertures are gated – they open and close on demand. And they need to open at the right time and let the right ion through. Like so much in biology, the details are complicated, but opening the gate seems to be activated by a cascade of protons initiated by a trigger proton that quantum-mechanically tunnels through a thin barrier to start the process off.*

There are also claims that neurons, like many living cells, emit photons with low intensity, at different wavelengths, perhaps correlated with different chemical processes going on in the cell. Could these photons constitute a parallel information network operating alongside the standard electro-chemical one? Brain cells might behave like biological lasers, sending photons across synapses, between neighbouring neurons, or over large distances along axons that serve as optical waveguides. This would offer a high-fidelity signalling mechanism for controlling the brain's electrical activity, analogous to the way that laser instructions along optical fibres can be used to control the movement of electric trains.

The most speculative suggestion is that quantum coherence

---

* It is important not to conflate the possible existence of quantum effects in the brain with the thorny issue of whether consciousness plays a fundamental physical role in quantum mechanics, a topic that will be discussed in Chapter 11.

can be maintained over large distances, to coordinate neural activity among different parts of the brain. A well-known and puzzling feature of brain activity is its synchronization across the cerebral cortex. In neurodegenerative diseases such as Parkinson's, this synchronization falters. Yong-Cong Chen and colleagues at the Shanghai Center for Quantitative Life Sciences have proposed a specific mechanism of quantum synchronization involving the myelin sheaths that surround axons.[5] The physical conditions in these sheaths favour the emission of pairs of entangled photons. The researchers speculate that 'quantum entanglement will effectively synchronize neuronal activity throughout the brain, shedding light on the "synchronization" puzzle.'

Even if these claims are wide of the mark, they may contain a germ of truth. It would then be of great interest to quantum engineers to determine how biological tissues can maintain quantum coherence or serve as photon waveguides. Cells are complex labyrinths replete with organized structures honed by billions of years of evolution for efficient functionality. They contain a wide variety of materials and structures. If there is a quantum advantage to be found, it is likely that some types of cells would have discovered it. There may be specialized components that could be harvested and re-engineered for quantum technology, or mimicked for use in quantum sensors or even in quantum computers. All told, it seems as if the nascent field of quantum neuroscience has the potential to lead to groundbreaking advances in cognitive enhancement, sensor technology and possibly fundamental physics too.

Quantum vitalism?

The burning question is this. Has life merely stumbled across a few quantum quirks along the evolutionary way, or are living organisms irreducibly quantum systems in their essential nature? Living things are the most complex systems we know. Among the myriad nooks and crannies of organisms there is plenty of scope for quantum mechanics to work its transformative magic – some tunnelling here, entanglement there, superpositions inveigling themselves into the molecular interstices, facilitating, speeding, amplifying, linking spookily across the teeming microstructures. The rich fabric of biology might just be suffused with some sort of quantum fairy dust, a latter-day life force. It is a beguiling vision, but one that is fiendishly hard to test. It is not just the complexity – the sheer difficulty of identifying the subtle quantum activity within the molecular mêlée – it is that the general physical conditions for life seem to be inimical to sustained quantum activity. The pervasive dampener of decoherence, boosted by the warm and wet conditions that characterize biology's comfort zone, serves to destroy the fragile quantum states necessary for anything very magical to be accomplished. Recall the extraordinary efforts that physicists have to go to in order to build a quantum computer, or even demonstrate many routine quantum phenomena – temperatures near absolute zero, isolation from environmental disturbances, carefully designed probes. Life, by definition, is an open system operating far from thermodynamic equilibrium, exchanging energy, heat and materials with its surroundings, and depending on liquid water. These requirements sound like the perfect storm of a quantum-destroying decohering environment. And yet, as I described in the previous sections, life has found a way, at least here and there, to harness genuinely non-trivial quantum effects to improve its prospects.

I have dabbled in quantum biology for over twenty years. I remain fascinated by the idea that perhaps some aspect of quantum mechanics is the key to understanding life, but I have yet to see any real evidence for it. That could be because living matter is so complex and quantum effects happen so fast that we don't spot the give-away signs. Perhaps certain distinctive patterns of quantum coherence permeate living cells, or even entire organisms, and we cannot yet detect this activity, let alone decode it. But in fifty years it may be a different story. It's possible we might by then have a type of 'life meter' that could detect and analyse highly dynamic quantum patterns in simple life forms.

Personally, I think there is a basic and rather straightforward problem about quantum mechanics and biology, quite apart from the 'perfect storm' of decohering conditions that pervade living matter. It boils down to this. Quantum mechanics is what physicists call a linear theory. In simple terms, in a multi-component linear system the whole is merely the sum of its parts. You see linearity displayed in quantum superpositions – you just add the parts of the wave function together in a two-plus-two-makes-four manner. And you can take the components apart again to get back what you started out with. That is to say, quantum mechanics is time symmetric. Apart from a small mathematical detail, Schrödinger's equation is unchanged by time reversal; if you evolve a wave function forward in time, there is an allowed physical process that evolves it back again. In fact, doing and undoing quantum transformations is standard quantum technology practice. By contrast, biosystems are highly *non*-linear. Suppose you have one type of organic molecule with a function A and another with function B, then combining the two molecules usually won't result in a function $A + B$, but a totally new function C; we might call this adding two and two to get five: the wholes are much more than the mere sums of the parts. And biosystems obey equations that are distinctly *not* time reversible. The bottom line is this. I don't think you can explain a highly non-linear physical system with linear processes.

But . . . here is where it gets intriguing. The linearity and time reversibility in quantum mechanics breaks down at the point of measurement. Measurement is irreversible. When the quantum system is coupled to a macroscopic device like a meter, the combined system is now non-linear because the internal workings of real laboratory equipment obey non-linear equations, as do organisms. All living matter constitutes a boisterous 'measurement environment' in which any quantum process is situated. Quantum mechanics is most relevant to biology, not at the stage where the components change in a linear way, but at the measurement step – where the quantum world (somehow) transitions to the classical. This, I believe, is where Schrödinger's 'new kind of physical law', mooted in his celebrated book *What Is Life?*, comes into play. It is precisely at the edge of life and the edge of classicality, where physics, chemistry and agency intersect, that we might find a new kind of physical law operating.

PART THREE

Cosmic Speculations

CHAPTER 8

The Myth of the Void

On the floor of Westminster Abbey, not far from the plaque emblazoned with Dirac's equation, lies a second tribute to quantum mechanics, this time in commemoration of the late Stephen Hawking, who died in 2018. Hawking became an international celebrity in 1974 when he claimed that black holes are not black after all, but glow with heat radiation. He came to this startling conclusion by applying the theory of quantum mechanics to these monstrous objects. The Abbey headstone displays the formula that Hawking derived, and you can see Planck's constant h right there on the slab.

When quantum mechanics was formulated a century ago, the founders were not especially preoccupied with technology. They understood that the conceptual transformation they were uncovering went far beyond practical applications: it would require nothing less than a thoroughgoing reconceptualization of how the physical universe works. Although the initial flurry of activity focused on the properties of atoms and molecules, by the 1930s, following Edwin Hubble's announcement that the universe is expanding, quantum physicists began extending their ideas to astrophysics and cosmology. Dirac, for example, believed he had found a link between the age of the universe and the speed of atomic processes. Schrödinger suggested that the expansion of the universe alone might serve to create photons. But it took another half century before these speculations could be properly investigated.

The sticking point in the 1930s was a deep confusion concerning the implications of quantum mechanics for empty space. Long ago, Aristotle proclaimed that 'Nature abhors a vacuum.' And he was right. Democritus had introduced the concept of atoms that moved in a void, sticking together to make matter. What we now call atoms are very far from the inert indestructible specks that Democritus proposed. But the same is true of the void. According to quantum mechanics, true emptiness is a fiction. We now know that even a total vacuum is a seething ferment of activity in which all manner of particles spontaneously spring into being uncaused, coaxed out of nowhere by quantum magic. They live only fleetingly before vanishing again, flickering on the edge of reality, their erstwhile existence insubstantial and ephemeral. Black holes, Hawking discovered, disrupt the quantum vacuum and turn that evanescent activity into a flux of heat radiation. The predicted effect is small, and to date, it has never been observed, but most physicists are convinced that 'Hawking radiation' from black holes is a real phenomenon. Armed with this result, cosmologists began to suspect that quantum mechanics might require us to rewrite the entire story of the universe.

To understand these momentous consequences, let's start with a careful look at the strange but crucial concept of the quantum vacuum.

Much ado about nothing

The quantum story began when Max Planck pondered the nature of heat radiation, and suggested that electromagnetic energy comes in little packets, or quanta: recall the discussion about the kiln on p. 10. Each electromagnetic wave in the kiln might have 0, 1, 2, 3 . . . quanta of energy. Here I want to focus on 0, that is, no quanta present: just emptiness – a vacuum. (I'm restricting this to photons for now, ignoring all other particles.) To keep it simple, imagine the kiln to be a large box with perfectly reflecting walls. Any radiant heat, or light, in the box can be envisaged as vibrations of the electromagnetic field, quivering from one end of the box to another, reflecting off the sides. (Think of something like a cup of coffee on a table with a juddering spin dryer going flat-out nearby; the surface of the coffee shudders in a standing wave pattern.) A photon is a quantized energy level of *one* of these vibrations. To say there are no photons in the box means that every wavelength is in the lowest allowed energy state of the vibrations.

That means no vibrations at all, right? Wrong! To understand why, consider a plucked guitar string. Envisage grasping the string in the middle and pulling it to one side. It requires some work to deform the shape. The energy you expend isn't lost, but gets stored in the string as elastic energy – an example of what is called 'potential energy' because it has the potential to do work. In the case of the guitar, 'work' means generating sound, which happens when you let go of the string and let it vibrate freely. The elastic potential energy gets converted to energy of string motion, i.e. kinetic energy. When the string passes the mid-point, it is straight. At that instant, the speed – hence kinetic energy – is at a maximum and the potential energy is momentarily zero. Then the string over-shoots and deforms in the other direction – kinetic energy gets converted back to potential energy. And so on, back and forth, as the string oscillates with a certain frequency. Quantum mechanics demands that the energy of the oscillations must be 'quantized', that is, it can be increased or decreased in little jumps only. For a real guitar, the jumps are far too small compared to the total energy of the vibrating string for us to notice, though it's possible to directly observe quantized vibrations in molecular-sized 'guitar' strings, such as carbon fibres stretched between two anchors.

Now we get to the key point. Surely an *unplucked* string, sitting idle, will have precisely zero energy? The answer is no. For the energy to be zero, the string would have to be both stationary (no kinetic energy) and straight (no potential energy) at the same time. Is that possible? Not according to Heisenberg's uncertainty principle, which requires that you can't know *both* the position and the motion of something at once. Let's apply this to a guitar string. How can you determine if it really is straight? Well, you could check by measuring the position of its centre. The uncertainty principle then says the motion of the string cannot be known. Suppose you make a different measurement to check that the string isn't moving? The uncertainty principle says that you cannot also know where that part of the string is located. There is an inescapable trade-off between checking that the string is straight and checking that it is at rest. More precisely, there is a trade-off between knowing the potential energy and knowing the kinetic energy. You cannot know both exactly, at the same time. A detailed calculation shows that this fundamental fuzziness in the two forms of energy implies that the string can never have *precisely* zero energy. Even in the ground state – the lowest energy state – there is an irreducible residue called the zero-point energy (a misleading term, since the energy isn't zero) that can never be removed. How much? It is equivalent to *one-half* of a single quantum of energy.

Let's make the leap from guitars to light. Both are vibrations of a sort, and the same basic quantum physics applies. The quantum theory of oscillators requires that each frequency will have a zero-point energy of one-half the energy of a photon of that frequency. And that is true of every frequency separately. What this means is that even when there are no photons around – a vacuum – there is still a residue of electromagnetic energy present deriving from the zero-point quantum energy of *all possible* electromagnetic waves.

There's an alternative way of thinking about the quantum vacuum energy which is very popular. A version of Heisenberg's uncertainty principle says that just as there is a trade-off between position and momentum, so there's a trade-off between energy and time. That is, the amount of energy present in a quantum system is not well defined for short periods of time: you can't precisely measure energy and duration together. Loosely speaking, this means that a photon can pop into existence from nowhere, using 'borrowed Heisenberg energy', so long as it vanishes again within the strictly prescribed time limit. (For the curious, a virtual photon of visible light would survive for a mere ten-thousandth of a trillionth of a second.) These fleeting photons-from-nowhere are called *virtual* particles to distinguish them from 'real' photons that are emitted by atoms, for example in this alternative picture, we can say that the vacuum is teeming with virtual photons that bubble up from nowhere and promptly vanish again. And by extension, there will be all manner of other virtual particles present too.

Which picture of the quantum vacuum is the more accurate: the one with the waves, each of which has an irreducible zero-point energy, or the one with a throng of virtual particles coming and going? They are both right in their own way, and simply reflect the fundamental wave-particle duality inherent in quantum mechanics.

So how much zero-point energy is there in the quantum vacuum in total? Embarrassingly, the theory gives an absurd answer – infinity. Each frequency has its own zero-point energy and there is no limit on the number of frequencies in an electromagnetic field. Even worse, the zero-point energy scales with the frequency, so it just keeps going up as you consider higher and higher frequencies. The existence of runaway zero-point energy is a reprise of the runaway heat problem of classical physics – the ultraviolet catastrophe – which I described on p. 10. But in the present case, there is no heat – the quantum vacuum is the absolute zero of temperature. Yet there is still energy, apparently limitless. The runaway heat problem was solved by introducing the quantum of energy, but now the theory is *already* quantum, so that's no help here.

Can the quantum vacuum be detected?

An absurd answer might suggest a nonsensical theory – which greatly troubled physicists in the 1930s – but before we rush to throw out quantum mechanics, we need to examine an escape clause. A prediction of infinite energy is undoubtedly absurd, but in most of physics energy per se is unmeasurable; only energy *differences* matter. For example, the energy needed to raise a lift from the ground floor to the sixth floor of a hotel is the same whether the hotel is next to the sea shore or up on the Tibetan plateau. (Actually, it will be very slightly less in Tibet because gravity is a bit weaker at altitude, but that's beside the point I am trying to make.) With that let-out, we could simply re-define the energy of the vacuum to be zero and forget about the inconvenient infinity.

Except it's not that simple. The quantum vacuum is an all-pervasive plenum – a sea of energy hidden from casual inspection. But it's there alright, lurking all around us (and inside us). It can't be ignored because there are circumstances where it can be disrupted and its existence physically manifested. I already mentioned black holes, but they were unknown in the 1930s, so the first such situation to be studied was the vacuum energy around and inside an atom. (Atoms aren't solid objects, of course, but consist of dot-like electrons whirling around a compact nucleus.) In the presence of matter, virtual photons of the quantum vacuum are subject to the same laws of optics as real photons – phenomena like refraction and reflection. And conversely, matter is affected by the restless activity of the quantum vacuum. For instance, atomic electrons can sense all those clamouring virtual photons, and as a result their energy levels get very slightly shifted. The magnitude of the shift can be calculated, and indeed was calculated in the 1940s. An experiment performed in 1947 confirmed a 0.1 per cent change in the difference between two energy levels of the hydrogen atom caused by the quantum vacuum activity. This may seem a trifle, which in a sense it is, though it was a triumph of theory and experiment, which concurred to phenomenal accuracy. But it served to demonstrate that the quantum vacuum has to be taken seriously and has measurable physical effects. And as we shall see, that tiny atomic energy level shift turned out to reveal the tip of a very big iceberg.

Another consequence of the quantum vacuum is the existence of a short-ranged force between electrically neutral atoms. Such a force had been proposed by Dutch physicist Johannes Diderik van der Waals in 1873 to account for the properties of gases, but he offered no explanation for the origin of the force. It took nearly eighty years for van der Waals's proposal to be explained as an effect of the quantum vacuum. Here is how it works. If you apply an electric field to an atom, it pulls the positive charges one way and the negative charges the other. This lopsidedness induces a corresponding electric field within the atom (the atom is said to be 'polarized'). An atom immersed in the quantum vacuum is subject to the ceaseless turmoil of fluctuating electric fields associated with the virtual photons, which then bestow upon the atom a fluctuating electric field (polarization) of its own. If another atom is located nearby, the two induced fluctuating fields will interact, causing a small, short-ranged force of electric attraction between the atoms, even when they are overall electrically neutral. These are the very forces that van der Waals guessed would be there all that time ago. Curiously, they show up in the living world too. A well-known example concerns geckos. These are lizards that have an amazing ability to cling to walls and ceilings. Geckos' feet have millions of tiny hairs as small as a few hundred nanometres in radius. Because of these hairs, the gecko is capable of holding its weight with only one toe of one foot. Geckos remain aloft, care of the quantum vacuum!

A more direct demonstration of quantum vacuum forces was suggested by the Dutch physicist Hendrik Casimir in the late 1940s, so it goes by the name of the Casimir effect. It is manifested as a tiny force of attraction between two parallel reflecting metal plates (mirrors), even though they are electrically neutral and situated in a complete vacuum (see Fig. 14). The explanation comes from the fact that the plates screen out some of the quantum vacuum energy. There will always be a longest electromagnetic wave that will just fit into the gap between the plates (like the lowest note on a guitar string). Waves with a greater wavelength, which might be possible in unbounded space, cannot fit into the gap; they are excluded. And so is their zero-point energy. Smaller gaps exclude more waves, and so exclude more zero-point energy. Thus, there will be an energy difference between small gaps and big gaps, or between narrowly spaced and widely spaced plates.

Why does the absence of some vacuum energy between the plates imply a force of attraction between them? The easiest way to think about it is to note that if someone pulls the plates wide apart, the vacuum energy in the space between the plates will go up (because more waves can fit into the gap). Where does that energy come from? It can be traced to the person manipulating the plates. Whenever someone moves an object against a pulling force, it is necessary to do work, which means expend energy. The Casimir force can be calculated precisely from quantum mechanics, and sure enough, when the work done to separate the plates

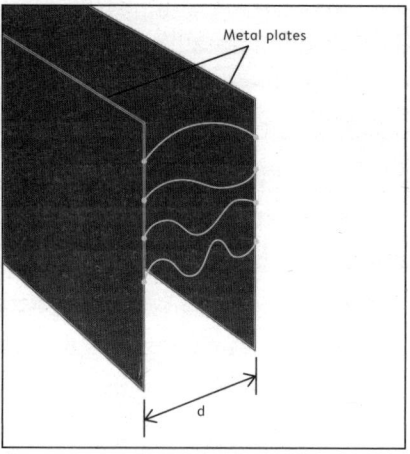

Figure 14

Casimir effect. The two parallel reflecting metal plates a distance *d* apart permit only certain discrete wavelengths of the electromagnetic field to exist between them. The total zero-point energy in the gap between the plates is therefore reduced on account of the absence of the excluded waves. That energy deficit produces a tiny force of attraction between the plates.

against that force is calculated, it corresponds exactly to the additional energy that appears between the plates.

The Casimir effect is sometimes described as 'the force from nothing', but of course the quantum vacuum is far from nothing, and Casimir set out to measure his eponymous force with the help of his colleague Dirk Polder. However, this was long before the advent of nanotechnology that enables delicate forces to be detected on a molecular scale. Over the years, many experiments have been performed with increasing precision to confirm the reality of the Casimir effect. One significant improvement was made in 1997 by physicist Steven Lamoreaux,[1] who used a type of pendulum which would move when influenced by extremely small forces. Recently, experiments have gone beyond the original parallel-plate configuration to explore quantum vacuum forces for different surface geometries – spheres, cylinders and other shapes.

A flash in the dark

New possibilities open up if Casimir's mirrors are allowed to move towards or away from each other. Now the energy of the vacuum in the gap will change with time. What is the result? This problem was considered theoretically as long ago as the 1970s, and it was immediately clear from the mathematics that the moving mirrors could create photons out of the vacuum – just like a moving electric charge does, only in this case the mirrors are electrically neutral. Nevertheless, they produce light when they move. In point of fact, you don't need two mirrors to jiggle up the quantum vacuum; a single accelerating mirror will do. The creation of light by a moving mirror has a ready interpretation in terms of virtual photons. Imagine such a photon that pops into existence briefly, but before it can do its usual rapid vanishing act, it gets struck by the mirror. When a real photon hits an approaching mirror, the reflected photon is blue-shifted relative to the incoming one; that is, it gains energy. The accelerating mirror likewise boosts the energy of any virtual photon impinging on it, maybe by enough to promote it into a real photon. The mirror uses some of its acceleration energy to 'pay off the virtual photon's Heisenberg debt' (see p. 169), thus sparing the photon the obligation of disappearing again. The result is a flash of light that illuminates the darkness of the quantum vacuum.

I worked on a wide range of problems involving the quantum physics of moving mirrors in the mid-1970s with colleagues Stephen Fulling, Lawrence Ford and William Unruh, and several students. We pretty much worked out all the basics. Disappointingly, the actual numbers looked pretty pathetic unless the mirror could be exceedingly rapidly accelerated to close to the speed of light. The flux of photons from realistic accelerating mirrors was pitifully small and, it seemed to us at the time, highly unlikely ever to be observed in an experiment. Over the years, though, there have been many ingenious proposals for glimpsing or amplifying moving mirror radiation, with little in the way of actual results so far. But a couple of recent experiments come close to the spirit of our original calculations. The mirror is replaced by a superconducting circuit that can undergo superfast changes, resulting in (real) photons being created from the quantum vacuum. In one of these experiments, Pasi Lähteenmäki at Aalto University in Finland and his colleagues cooled an array of 250 SQUIDs (see p. 142) to within 50 thousandths of a degree above absolute zero, so cold that it became pretty much a quantum vacuum. They then used magnetic fields to very rapidly alter the characteristics of the superconductors in a way that changed the effective speed that light propagates through the space between them. Although not quite the same as a moving mirror, it had a very similar effect, turning virtual photons into real pairs of entangled photons. In another experiment, Christopher Wilson and colleagues, then at Chalmers University in Sweden, also used abrupt changes to a SQUID, to alter the effective length of a waveguide – a channel that corrals electromagnetic energy – mimicking the way that the channel would behave if it terminated in a mirror that suddenly moved. Again, they were able to detect real photons coming out of the quantum vacuum.[2] It's hard to think of any immediate technological uses of moving mirror radiation, but one never knows!

A far more dramatic example of quantum vacuum disruption is Hawking's black hole radiation. One way to think about it – the way that Hawking himself favoured – is that virtual photons get born in pairs. If a pair pops up just outside the surface of the black hole (the so-called event horizon), it will be in a precarious location. Suppose, during its transitory existence, one of the virtual photons ventures too close to the black hole? Then, wham, the black hole grabs it before it can dissolve back to nothing, leaving the unaccompanied twin to fly away. The black hole's gravitational energy pays off the survivor's Heisenberg debt, permitting it to speed off into the universe, effectively immortalized care of its sacrificed partner. Expressed graphically, the gravity of the black hole is so ferocious, it rips apart the ghostly fabric of the quantum vacuum, promoting its fleeting virtual inhabitants into real particles. A black hole is therefore a machine that turns the 'nothing' of the quantum vacuum into something tangible – a glow of heat radiation.

Dreams of cosmic engineering

One of the oddities we found about moving mirror radiation is that, depending on the exact mirror trajectory, the flux of energy that an accelerating mirror produces can be *negative*, by which I mean the energy it spews forth will be *less* than the energy that is already there in the quantum vacuum. This was an exciting discovery! What engineering possibilities might arise if one had a controlled beam of negative energy at one's disposal? Would it be like a 'beam of dark', cancelling light? Or a beam of cold, annihilating heat?

As energy is equivalent to mass, negative energy implies negative mass, which in turn implies antigravity. For example, the vacuum energy between the metal plates in the Casimir effect is negative; however, it's negligible compared to the mass of the plates, so the total mass (energy) of the Casimir system is always positive. But with an unrestricted flux of negative energy at your command, you could, it seems, direct the beam into a metal box until enough negative energy had accumulated so that the total mass of the box plus its quantum contents would be negative overall. The box would then experience a negative gravitational force; it would levitate! The Earth's gravity would fling the box up into space, like the fictional substance called 'cavorite' in H. G. Wells's story *The First Men in the Moon*.

Further dramatic effects could be achieved by directing a

negative energy beam at a black hole, a scenario identified long ago by Lawrence Ford.[3] Black holes form when a star implodes under the gravity of its immense weight. In a simple model, the material of the star would shrink to a point of infinite density, known as a singularity. A singularity is like a tear or rip in spacetime – a boundary where space and time cease to exist. All known physics would break down there. If such a thing were to form in plain sight – a so-called naked singularity – then there would be a suspension of the natural order even at the classical level because something that didn't have prior existence in the universe could emerge from the singularity. It would be an event without a cause on a cosmic scale. This alarming prospect led Penrose to propose what he called the cosmic censorship hypothesis – that singularities which form by gravitational collapse cannot be naked, but will always be decently shielded from our gaze by a black hole. The surface of a black hole is called an event horizon precisely because no events inside it, including those at the singularity, can ever be witnessed by an observer on the outside, on account of the fact that even light cannot escape the gravitational prison. However, nobody has proved cosmic censorship, so the question remains open as to whether naked singularities may be formed somehow.

Ford thought he had discovered a way to do it by directing a sustained beam of negative energy at a suitably prepared black hole. Theoretically, it could then be converted into a naked singularity. By 'suitably prepared' I mean electrically charged. The mathematical description of a charged black hole is very simple and has a surprising algebraic feature: if the charge to mass ratio of the black hole were to reach a certain critical value, the event horizon would abruptly vanish, according to the algebra, revealing the dreaded singularity to the outside universe. Suppose, then, a black hole had an electric charge very close to the critical value. We can imagine an inquisitive supercivilization charging up a black hole to the limit as an experiment. They couldn't press on and add more charge to make a naked singularity because the electric repulsion from the black hole stops anyone topping up the charge beyond the tipping point. But with Ford's strategy it is a different story. Imagine a beam of negative energy directed at this critically charged object, which is teetering on the edge of cosmic disaster. The flux from the beam would reduce the total mass of the object while leaving the electric charge unchanged, and so would tip it over the edge, driving the ratio of charge to mass above the critical value. At that point, the horizon would vanish, exposing the feared singularity to public view!

Creating a naked singularity is not the only (admittedly fanciful) example of cosmic engineering mischief that hinges on the manipulation of negative energy. Another is the wormhole in space. A wormhole is rather like a black hole, but with a crucial difference. Whereas falling into a black hole would be a one-way journey to nowhere – a black hole has an entrance but no exit – you could fall through a wormhole and come out somewhere else, maybe a long way away. It would be a sort of tube of space that links distant places to form a shortcut – an alternative route from A to B that bypasses normal space. Like black holes, wormholes are a sci-fi favourite – but whereas black holes are real, wormholes remain a theoretical idea only. So why are we bothered by the mere prospect of a wormhole? Well, it turns out that a wormhole could be used not only to travel through space, but to travel through time too. With some simple manipulations, a traversable wormhole could be turned into a time machine.[4] An object (e.g. a person) falling through the wormhole in one direction would be transported to the future, while falling the other way it (or they) would be projected back into the past. Fascinating though such a device might appear, it is regarded by most physicists to be equally as abhorrent as a naked singularity, on account of the causal chaos that would ensue. Travelling back in time is fraught with paradoxes, such as the time traveller killing their younger self. Which is what makes time travel so engaging in fiction, but so disturbing to physicists.

The conundrum here is that, even if wormholes don't exist in nature, a sufficiently advanced technological community might be able to make one, perhaps as part of a cosmic transport network, as envisaged by Carl Sagan in his sci-fi novel *Contact*. This is where negative energy comes in. According to standard gravitational theory, if a wormhole were to form somehow from normal matter (a collapsing star, say), it wouldn't last long because it would almost immediately shrivel and pinch off to nothing before anything could get through it and exit the other end. But if the 'throat' of the wormhole contained, not star-stuff, but *negative energy*, with its associated antigravity, it could be propped open long enough for something – or someone – to pass right through. Whether or not traversable wormholes can exist in principle thus hinges on identifying a source of negative energy. Well, we know of one: the region between the plates in the Casimir effect.

The foregoing scenarios are not intended to be practical suggestions. Rather, they fall into the time-honoured category of 'thought experiments', which were used to such decisive effect by Einstein and others at the birth of quantum mechanics. To be sure, in the latter case, the thought experiments became real experiments within a few decades, but it would be rash to suppose that our twenty-second-century descendants will be creating naked singularities or wormholes in space. Nevertheless, when the name of the game is to ponder the implications of negative energy for the rational order of the cosmos, these sorts of hypothetical deliberations are extremely valuable, and might lead to the discovery of new physics. Which is why the scenarios I have been describing have been published in respectable scientific journals.

In that spirit, what can we then conclude from the foregoing thought experiments? Clearly, the control of substantial amounts of negative energy threatens to unleash some unpalatable, even paradoxical, consequences for the physical universe. But haven't I already claimed that negative quantum energy exists, albeit feeble in quantity? That is true. But like everything in quantum physics, you have to examine the theory's small print. For example, to make negative energy with a moving mirror, the mirror has to accelerate in the same direction as the beam it creates, so the mirror eventually crashes into the metal box or charged black hole, and puts a stop to the fun. When you do the sums and add up the total amount of negative energy, it is too small to do much harm. For example, the horizon of a critically charged black hole merely scintillates on the edge of existence – a phenomenon that Ford dubbed 'cosmic flashing'. It never disappears for long enough for anything to emerge from the singularity and invade the universe.

How about the wormhole? Again, there is a problem. Even if a way could be found to infuse the throat with enough negative energy for it to remain open, the ensuing formation of time loops would likely cause the quantum vacuum to explode. A time loop means that an object can loop back into its own past and become its earlier self. Suppose that object is a virtual photon in the quantum vacuum. Virtual photons live on borrowed time, remember, care of the Heisenberg energy-time uncertainty relation (see p. 169). But a time loop can take a virtual photon whose time is up right back to the moment it was born: the borrowed energy wouldn't have to be repaid after all! The virtual photon would be in a state of suspended animation. The entire cohort of 'un-dead' virtual photons would swarm around the wormhole, surging with unbounded borrowed quantum energy, generating a gravitational field that swamps the negative vacuum energy invoked in the first place to prop the wormhole open. It seems very likely (but is not definitively proved) that this runaway quantum vacuum energy would always serve to destroy the wormhole as soon as the threat of a time loop arose.

All these caveats undermine the prospects of an advanced civilization using negative energy for astro-engineering purposes. The story always seems to be the same. Whatever scenario one may cook up to use negative energy to bring about something exciting or alarming, nature always counters with a compensatory effect to frustrate the attempt. Which raises a very deep question about the way the universe is put together. How is it that quantum mechanics, while flirting with negative energy, nevertheless circumscribes it so tightly that it doesn't unleash bizarre possibilities, like time loops? Striking though these let-out clauses seem when investigated, there is no general theorem or argument that it must be so in all circumstances. And after all, why should quantum mechanics care about crazy spacetimes anyway? Quantum mechanics doesn't know anything about gravitation. It has no obligation to save the universe from absurd scenarios such as naked singularities. And yet as far as it has been investigated, the laws of quantum physics always seem to rescue the universe from the brink of serious unpleasantness. Which in my mind hints at a deep linkage between these two branches of physics – gravitation and quantum mechanics – that so far hasn't come out in any straightforward attempts to marry them.

Vacuum-powered spacecraft?

The discovery that the vacuum is full of quantum energy, even if the forces it exerts are tiny, has proved irresistible to some starry-eyed engineers. Could we not tap the quantum vacuum as a source of energy to drive a spacecraft, for example? Why take rocket fuel from Earth when empty space itself is replete with virtual photon energy for the taking? Just tank-up on the way to the stars! This beguiling idea might have been behind NASA's Breakthrough Propulsion Physics programme announced in 1996, with the purpose of discovering a means of propulsion 'that requires no propellant mass, propulsion that attains the maximum transit speeds physically possible, and breakthrough methods of energy production to power such devices' . . . such as warp drives, wormholes and, yes . . . 'vacuum fluctuation energy'.[5]

The fundamental snag with this idea is that, as I have mentioned, only energy *differences* are relevant to mechanical effects. However much energy the vacuum may contain theoretically, it is uniformly spread across empty space, a featureless sea of quantum activity. Extracting it to run any sort of motor requires coupling the device to a region with *lower* energy to permit the quantum vacuum energy to flow from higher to lower. It's the same principle with hydroelectricity. The lake behind a dam may represent a huge amount of potential energy, but you have to let water flow out to tap it. Now it's true that material objects can disturb the quantum vacuum: they create small regions of higher or lower vacuum energy. That allows mechanical work to be done, for example, by letting the Casimir plates draw closer together. But the energy extracted isn't unlimited. Once the plates touch and stick together, that's it. To repeat the process, you have to pull the plates apart again, which will require the same amount of energy as you extracted, leaving you with no net gain. Another problem is that, in the absence of nearby matter, the vacuum energy is the same in all directions. To propel a spacecraft to a target destination, the star Alpha Centauri say, something has to break the symmetry and shove the craft that particular way. Small wonder then, when the NASA programme concluded in 2003, the website carried the deflating statement: 'No breakthroughs appear imminent.'

Although powering a spacecraft from the quantum vacuum in mid-flight appears to be a lost cause, it is possible to use quantum teleportation to transfer tiny amounts of quantum vacuum energy from place to place by once again exploiting entanglement. The earlier discussion of teleportation (p. 94) used pairs of entangled photons. But even when there are zero photons present there is still a form of entanglement in the vacuum state itself, between different regions of space. It arises because the quantum vacuum is defined by considering the quantization of waves (modes) in the electromagnetic field that are spread out across the entire space (see p. 167). A clever way to transfer energy using vacuum entanglement was found by Masahiro Hotta, a theoretical physicist at Tohoku University in Japan,[6] and it works like this. Alice would like to pluck some energy from the quantum vacuum in her lab, but because the quantum fluctuations in energy are random, any attempt to grab energy from them with a gizmo would be a literal gamble – Alice would lose as often as she won, and the average gain would be precisely zero. But if Alice has an accomplice, Bob, across town, then there is a little manoeuvre they can employ. Bob can perform a quantum experiment in his lab, expending some energy to obtain information about nearby quantum vacuum fluctuations. Then he calls Alice with the details. Alice, knowing that Bob's vacuum and hers are entangled, can then perform a carefully arranged measurement on her vacuum using Bob's information and – hey presto! – some vacuum energy gets harvested in Alice's lab. No energy has been created: rather, Bob has merely told Alice when her quantum vacuum will fluctuate to her energy advantage. (In fact, some energy has been expended in making the phone call.) Vacuum energy teleportation, now demonstrated, could find applications in quantum refrigeration and other technological systems such as the quantum internet (see p. 110), but is unlikely to be of service as rocket fuel.

Harnessing the quantum vacuum

Staying down to Earth, there are many other practical ways that negative quantum vacuum energy can be successfully exploited on a small scale.[7] The Casimir effect is one of them.

Although it is normally a very small attractive force, it increases sharply as the distance between the two plates decreases, so on the nanometre scale it can be hugely important. By way of illustration, two flat plates 10 nm apart feel an attraction equivalent of about one atmosphere pressure. As a result, Casimir forces are significant for micro- and nanotechnology. Very often they are a nuisance, because they cause components to stick together. The combination of sticking and friction has been dubbed 'stiction', and it is something of a headache for engineers. Recently there has been interest in designing nanotechnology components to either minimize quantum vacuum effects, or else to harness them as part of the mechanism. The strength of the force between two surfaces depends not only on their separation, but a variety of factors – the shapes of the surfaces, the material from which they are made and any fluid or other material in the gap between.

Casimir calculated his effect assuming perfectly reflecting flat plates, but metals are not ideal mirrors and total flatness is impossible. In more realistic cases, the Casimir force can vary significantly from a simple force of attraction as I have described it. In some circumstances, depending on the geometry and materials present, the force can be repulsive, leading to the possibility of a 'quantum buffer' to prevent components rubbing against each other, and opening the way to nano-machines with frictionless bearings. Sometimes the reflectivity of the surface can be abruptly altered, for example with a laser, thereby changing the strengths of the Casimir forces in a controllable way. The possibility of tailoring the geometry and materials enables engineers to design novel devices, offering potential for harnessing quantum vacuum effects in commercial products.

The quantum vacuum can even be manipulated to produce a rotational force, or torque, by creating a small gap between materials with oppositely twisted internal structure. Different materials can yield opposite effects, enabling a further level of control. Casimir forces also play a role in the human body: for example, red blood cells can stick together to form cylindrical stacks. Taking the Casimir effect into account will become crucial in the design of nanobots and other synthetic devices implanted into living organisms for therapeutic purposes.

You might imagine that if the vacuum is a sea of restless virtual photons, then a body travelling through it would be impeded in some way. However, this is not so. Although, as I have explained, an atom senses the vacuum fluctuations and experiences a slight shift in its energy levels as a result, it does not feel any frictional drag. The quantum vacuum has certain features of a pervasive medium, but viscosity is not one of them. That changes, however, if other material is present in the vicinity of the object of interest. Suppose the atom is situated close to a metal surface. Just as you can see an image of yourself in a metallic mirror, so an atom sees an image of itself behind the surface. Although the reflection is 'imaginary' (it is an image after all) it still has a real physical effect. If the atom is moving parallel to the surface, the image tags along, keeping pace with it. But here is where things get interesting. Assuming we are dealing with an ordinary metal like silver or gold or copper, and not a superconductor, the reflectivity isn't 100 per cent perfect. That's because light images (e.g. of your face in a mirror), and indeed all electromagnetic images, are produced by the movement of electrons in the metal. Silver, gold and copper have some electrical resistance, which causes the moving electrons to lose energy in the form of heat. So when the atom moves across, and close to, the metal surface, it has to drag its image with it, through the resisting medium of the metal. Although the quantum vacuum sea itself does not impede the atom, its image holds it back as the atom pulls it through the metallic mire. Obviously, this viscous force will depend on the specifics of the metal, the distance of the atom from it and so on. This phenomenon is sometimes described as quantum vacuum friction, but the friction is in the metal; the vacuum itself is merely a mediator. It's more like friction at a distance. The atom slows down in one place and the lost energy appears as heat in another.

Curiously, it is not the only case of friction at a distance. Something similar happens with the moon going around the Earth. The moon's gravity raises tides, not just of water, but of rock too, and the bulge travels round the planet as it spins. There is friction – think of the ocean tides crashing against the shore, for example – which slows down the travel of the bulge and causes heat in the Earth. Because the tidal bulge is held back somewhat, it won't precisely line up on the moon's position in the sky, but will lag a bit behind. The gravity of the bulge will therefore tug on the moon, serving to slow it slightly in its orbit. Over time, this effect causes the moon to drift further away (into a slower orbit) – by about 3 cm per year.

Not only can atoms reach across the gap to their images in metals, sound quanta (called phonons by analogy with photons) can do so too. Fans of the *Alien* movies will know that, in space, no one can hear you scream. But that's not quite true. As with all things quantum, there can be peculiar exceptions. If 'space' means the quantum vacuum, it turns out that sound can in fact jump the gap between two materials held very close. It's not that the sound waves themselves travel directly through the aethereal sea of virtual particles. Rather, the quantum vacuum fluctuations mediate an interaction between sound in one material and sound in the other, through the gap separating them. Sound, being vibrations of molecules, can couple to the vibrations of the vacuum and thence to their counterparts in the other material across the gap. And if sound can jump the gap, so can heat, which is just disorganized molecular motion. In a fascinating experiment reported in 2019, a collaboration between a group at the University of Hong Kong and the University of California, Berkeley, demonstrated quantum vacuum heat transfer in a nanoscale Casimir-type system.[8] The phenomenon can go alongside radiation and convection as modes of heat transfer, and while it is a tiny effect – too small for us to worry about thermos flasks – it could be significant in nanotechnology.

Whose vacuum is it anyway?

In the never-ending discourse concerning what is 'really there' in the quantum universe, nothing is so puzzling as nothing, by which I mean, how should a quantum physicist describe the state of 'no particles present'? As I have explained, this is called the vacuum state: a state that has no real particles of any sort, including photons, although it is replete with virtual particles. But the very concept of the quantum vacuum is a rather abstract one, beset with interpretational conundrums. To make matters worse, the quantum vacuum seems to have a certain subjective element to it.

In the summer of 1974, I was thinking about the lecture

by Stephen Hawking in which he announced that, because of quantum effects, black holes will glow faintly with heat radiation. I had been in the audience, and the mathematics baffled me. Frankly, I didn't believe Hawking's claim at the time. If black holes really did emit particles – thermal radiation – then I wanted to calculate it for myself, but with a simpler mathematical model than Hawking used. Fortunately, I had one to hand, because Stephen Fulling had already done most of the work and had sent me his calculations. Fulling had looked at quantum states as they might appear to an accelerating observer. Acceleration feels like gravity (think of how fairground contraptions can pull you down so you feel heavier, or give you the feeling of 'leaving your stomach behind'), and indeed, acceleration *is* gravity in a deep sense – an equivalence that goes back to Galileo. Furthermore, not only does an accelerating observer feel they are in a gravitational field, but, if the acceleration never stops, then the observer approaches the speed of light as judged by a stationary observer. And because no information can travel faster than light, that means there is a horizon in space beyond which the accelerating observer cannot see – just like the event horizon of a black hole. The upshot is that you can mimic the quantum theory of a field in the vicinity of a black hole by an analogous mathematical treatment for the much simpler accelerating observer system. So that's what I did. I soon found something deeply significant: what seems to be a quantum vacuum to a stationary observer looks like a bath of *heat radiation* to the accelerating observer.[9] More precisely, what to a stationary observer *is* a quantum vacuum, containing only virtual photons, appears to an accelerating observer to be populated by *real* photons.

Who was right? Are there particles present or not? An answer came shortly after, when Bill Unruh published a calculation in which he worked out what would happen to a particle detector (such as an atom) accelerating through what a stationary observer would swear is a quantum state that is totally devoid of any real photons. And sure enough, he found that the detector would click at just the rate corresponding to a bath of real photons with a spectrum identical to that of heat radiation. This simple analysis unleashed a decades-long debate about what is 'really going on'. Is the accelerated observer somehow mistaken? Are the particles that are registered in the detector phony? And where does the energy come from to excite the detector?

While there remain many nuanced positions on this topic, I myself have a rather straightforward view. I'm always tempted to think of a photon as a little blob of concentrated energy whizzing along. But when I put on my professional physicist's hat I know that won't do. Described mathematically, a photon is a quantum of energy assigned to a specific waveform – technically called 'a mode' – of the electromagnetic field, which, as I have already mentioned, is spread out over the whole space, just as a quantum of vibration in a guitar string is a property of the whole string. A single photon is a spatially extended entity. But extended how far? That depends on how the observer is moving. An eternally accelerated observer can see only a portion of the whole space, because a horizon blocks off parts of it. It's the same reason an external observer can't see inside a black hole. Thus, the modes of the field experienced by the accelerating observer are confined to a circumscribed region of space. A vacuum defined for modes in this restricted portion is mathematically not the same as a vacuum defined on the whole space – the one that seems empty to a stationary observer. Your quantum vacuum and my quantum vacuum can be different if we move differently. In the above case, your vacuum is my heat bath.

Where, then, does this leave the definition of a particle? Is the heat radiation detected by the accelerating observer 'really there' even when a stationary observer sees nothing? In my view, a 'particle' is simply what a particle detector detects, and that detector might be moving in all sorts of ways. Different detectors will register different results when exposed to the *same* quantum state. There is thus a subjective element, an observer-dependence, as to whether or not there are real particles in the observer's vicinity or whether the space is 'empty'. Which only goes to reinforce the fact that there is no such thing as a true void.

So is that it? Is the quantum vacuum just one of those compelling, intriguing, mind-bending quantum things that physicists love to contemplate, but whose effects are almost inconceivably tiny and of practical interest only in the field of nanotechnology? Not quite. The quantum vacuum bestows energy on empty space and there's an awful lot of space in the universe. Something that is tiny on a human scale might, cumulatively, make a huge difference on a cosmic scale. And indeed, that does seem to be the case. As we shall see, the invisible energy of the quantum vacuum, feeble though it is in the laboratory, may yet be the factor that determines the ultimate fate of the entire universe.

CHAPTER 9

Universe Out of Nothing

There can be no greater challenge to science than to explain how the universe came to exist. Cosmologists are convinced that the universe as we know it began with a big bang about 13.8 billion years ago, but naturally people ask why the big bang happened and what, if anything, preceded it. Did the universe somehow pop into existence out of nothing? If you've come this far, you will know that 'nothing' is not a word to be bandied about loosely. But, sidestepping the philosophical nuances for now, it's legitimate to ask whether quantum mechanics can explain the genesis of the entire cosmos – the ultimate mystery of mysteries. Was the big bang some sort of mega-scale quantum phenomenon, and if so, what can we say about it?

Quantum all the way up?

You might think it odd that I've turned from discussing objects on the atomic scale, where quantum effects dominate, to the subject of the whole universe. Within the scope of our instruments, which can reach over 13 billion light years into space, there is about a hundred trillion trillion trillion trillion tonnes of matter, which is about 80 powers of ten more massive than an atom. Can we seriously expect to successfully extrapolate quantum mechanics that far? Maybe not, but there's a popular argument to justify appealing to quantum effects at the time of the big bang at least. It's a simple numbers game.

As explained in Chapter 1 (p. 15), when Max Planck coined the term 'quantum' in 1899, he introduced a new constant of nature that bears his name. Its value gauges the smallness of quantum effects. For example, if you give me a specific frequency of light, I can multiply it by Planck's constant to tell you the energy of one photon possessing that frequency. Planck realized that by introducing a new constant into physics, he could combine it with other known constants – the speed of light and Newton's gravitational constant – to obtain natural units of length and time. These go by the name of the Planck length and Planck time: their values are approximately 10^{-33} cm and 10^{-43} s respectively. One can also use these constants to define the Planck energy, which is about 20 orders of magnitude greater than our best particle accelerators can achieve. Physicists agree that something seriously peculiar will happen in such a tiny bit of space or over a tiny interval of time, although they don't agree on what. It follows – in a general, handy-wavy sort of way – that when the universe was one Planck time old, some sort of quantum jiggery-pokery must have been going on.

There is another general motivation for applying quantum mechanics to the birth of the universe. The textbook description of the big bang is that it was the origin of space and time as well as matter. If there was no time before it, then the big bang would be an event without a cause. In classical physics, such a thing would be nonsensical, but quantum mechanics enshrines causeless activity in its very structure. Quantum mechanics is based on uncertainty and indeterminism: events are not fully determined by what preceded them. Think, for example, of a lump of uranium undergoing spontaneous radioactive decay. Ask why a particular nucleus decayed at 1.24 p.m. on a Thursday, and there is no answer. It just happened. So maybe the universe 'just happened' – a spontaneous, causeless quantum event that brought into being space, time and matter?

As it stands, these general arguments are not a lot of help, because we lack a decent theory of quantum cosmology to put flesh on the bare bones of this vague statement. But that's not for lack of trying. The idea of a 'wave function of the universe' was mooted by the physicist John Wheeler and his collaborator Bryce DeWitt in the 1960s. And by wave function of the universe, I mean a quantum description of everything there is – all matter and fields together with the gravitational field that describes the expansion of the universe itself. The entire package. Part of their motivation for this ambitious project was Wheeler's unease about the instant of cosmic creation that begot the expanding universe we inhabit. According to Einstein's general theory of relativity, as you run the great cosmic movie backwards, the universe shrinks faster and faster until everything is squashed at the beginning into a state of infinite density – a singularity – similar to what happens at the centre of a black hole (see p. 179), but in this case lying in our cosmic past. It's a moment where time itself begins – which sets many people's heads spinning. Wheeler thought that quantum fuzziness might smudge the initial singularity into something finite, implying that the big bang would not be the ultimate origin of physical existence, but just a highly dense, blurry state of space, time and matter that was presumably preceded by something else.

To get some idea of how such an analysis might work, focus on the simplest of all cosmic activity: the expansion of space. Imagine a vast spherical volume of space: because of the expansion, tomorrow this volume will be a bit bigger than today. What do you get if you describe this symmetric dilatory motion by a quantum wave function? The essence of wave functions is superposition: Fig. 3(a), for example, shows a wave packet for a particle in one dimension, which you could imagine moving to the right, as a travelling hump. In that case, the wave function is a superposition of many different positions for the particle and many different momenta. The cosmological counterpart to the particle's position might be the size of the universe, and the counterpart to the momentum, the rate of expansion. An analogous wave packet for the universe would then involve a superposition of many sizes and expansion rates. Unlike the particle, which can be in many different places in space, in quantum cosmology it is *space itself* that would be in a superposition. To make progress down that path, we have to think that there isn't just one expanding space but a plethora of them, blended together. In case you are wondering what on Earth that means, bear with me. I'm going to come back to it. In a fancier treatment, account would also be taken of how the expansion rate might vary in different places and different directions, and one would have to add all those possibilities to the superposition. Roughly speaking, the wave function of the universe will include the many possible shapes of space as well as different expansion rates, and presumably different topologies too.

Wheeler and DeWitt wrote down a formal mathematical equation for how the wave function of the universe would

behave – the Wheeler–DeWitt equation. It's the analogue of
the Schrödinger equation, but for the whole universe. Unfortunately, the equation can't really be solved in the standard
way, so it is often regarded as purely symbolic. Worse still, it
isn't at all clear what any such solution would actually mean.
After all, in laboratory physics, a quantum wave is an abstract
concept, promoted to concrete meaning only by a mathematical operation – the Born rule – that uses the wave function to
compute a set of probabilities for the outcomes of different
measurements (see p. 23). Defaulting for the moment to the
Copenhagen interpretation, the manifestation of these probabilities demands an external classical observer able to perform measurements. Since the universe is everything there is,
no such external observer exists. Furthermore, the very notion
of probability hinges on the idea of an ensemble – a collection
of identical systems. For example, you might have a box full of
identical coins randomly shaken. The probability of extracting three coins in a row all showing heads up is 1/8, four heads
1/16 and so on. But if there is only one universe, rather than
a box of them, what is the probability for this-or-that physical property supposed to mean? For now, I shall evade these
tough interpretational questions and go with the idea that
there is indeed a valid quantum description of the universe as
a whole, even if we aren't yet sure what it is.

Accepting, then, that the wave function of the universe
at least makes some sort of sense, we are immediately faced
with the problem of choosing an actual quantum state. In the
lab, the experimenter can prepare an initial state and then
see what happens. But who prepared the Great Cosmic Initial State? In the absence of some metaphysical assumption,

one can only hazard a guess at the mathematical form of this state. There is a general acknowledgement that, whatever the specifics, the universe started out simple, and is growing progressively more complex over time. Astronomical evidence strongly supports the idea that the universe was indeed highly uniform and featureless at the beginning. Nobody can say the universe *had* to begin that way, but it's as good a criterion as any to get started. In the early 1980s, Hawking and his collaborator James Hartle proposed just such a quantum state of primordial simplicity. The particular one they picked commended itself on mathematical grounds.

The Hartle–Hawking state has the attractive feature that the initial singularity is absent, as Wheeler long hoped, but in a curious way. What the Hartle–Hawking wave function describes is a universe without a singular first moment of time, but one in which time does not extend backwards for ever either. That sounds like an oxymoron. Either time begins or it doesn't, right? Well, Hartle and Hawking managed to pass between the horns of this dilemma because their wave function describes a universe in which (in a time-reversed sense) time 'turns into' space in the ultra-compressed circumstances of the big bang. And by 'turns into', I don't mean that one moment there are the familiar three dimensions of space and one of time, but a moment earlier there are four dimensions of space and no time. It isn't like flicking on a light switch: 'Let there be time!' Rather, the transition is gradual. The word gradual here is a euphemism, because the time scale that is relevant is the Planck time, which is 10^{-43} s – pretty fast by human standards. But there is still continuity. If you are wondering how to think about space 'turning into' time, remember that in quantum mechanics you have superpositions and uncertainty. The distance between two points in four-dimensional spacetime is a combination of spatial separation and temporal duration, and the mix of those can be fuzzed by quantum mechanics – a bit more space, a bit less time, superposed with a bit less space and a bit more time. It all comes out in the mathematical wash.

Remarkably, the general idea of a quantum origin of spacetime was mooted decades before the work of Hartle and Hawking, by a young Belgian priest and physicist, Georges Lemaître. Not only was Lemaître the first person to seriously advocate for what today we would call the big bang, he proposed that the originating mechanism would be quantum mechanical. In a paper in the journal *Nature* in 1931 entitled 'The Beginning of the World, from the Point of View of Quantum Theory', he wrote, 'The notions of space and time would altogether fail to have any meaning at the beginning . . . the beginning of the world happened a little before the beginning of space and time.'[1] Lemaître's allusion here to 'a little before' refers to the Planck time.

I have just given a flavour of what is an abstract and technical mathematical definition of the Hartle–Hawking state, and you would be right to be sceptical that any of this passes muster as a serious hypothesis for the coming-into-being of the entire cosmos. Many cosmologists are also wary. But as always when theory makes bold claims, there is a rush to ask for observational evidence. So, have astronomers spotted anything to support the claim that the birth of the universe was dominated by quantum mechanics?

Quantum fingerprints in the big bang's afterglow

When the universe erupted into existence with a bang 13.8 billion years ago, coughing out all manner of subatomic particles, it also glowed with intense heat. That primordial thermal radiation is still all around us today, slowly fading as the universe expands, so it now bathes the entire cosmos in inconspicuous microwaves. About 380,000 years after the big bang, the cooling cosmic gases ceased to glow and became transparent. Thereafter, the heat radiation was left almost undisturbed, so it offers us a snapshot of what the universe looked like at a mere 0.0028 per cent of its present age.

The cosmic microwave background has been intensively mapped by satellites and poured over by cosmologists for clues about the early universe. One of its most distinctive features is the slight variations – about one part in 100,000 – in temperature across the sky, representing hotter and cooler regions of the early universe. Variations in temperature track variations in the density of material, so the patchy distribution of heat indicates vast regions of slightly denser gases that were the precursors of galaxies and galaxy clusters. These 'seeds' of cosmic structure are thus a critical feature of the cosmos. But where did the slight but all-important density variations come from in the first place? A detailed statistical study of the temperature differences across the sky on different angular scales hints at an intriguing conclusion. The cosmic primordial heat radiation has been imprinted with variations that look suspiciously like quantum vacuum fluctuations, stretched by the expanding universe from atomic to cosmic dimensions. This has to be a huge clue and is compelling evidence that the universe was indeed born in some sort of quantum vacuum process.

It's possible to flesh out that very idea: that the large-scale structure of the universe is a frozen relic of primordial quantum vacuum fluctuations. I laid the groundwork for this explanation in the 1970s in collaboration with my student Timothy Bunch. The literature cites the Bunch–Davies quantum vacuum state as the starting point of this analysis. The favoured setting is the so-called inflationary universe scenario, in which the new-born cosmos, which might have started out in a complicated mess, underwent an enormous and accelerating growth spurt (dubbed inflation). Although inflation lasted but a split second, it was so powerful that it smoothed the universe out into a state of overall uniformity, embellished only by tiny quantum fluctuations etched into the structure of the universe, and which showed up 380,000 years later in the background distribution of heat radiation. Without those primordial quantum fluctuations there would be no galaxies, stars, planets, or people. So, in that respect, we have the quantum vacuum to thank for our very existence.

Do I believe in quantum cosmology? I'm conflicted. On the one hand, I like the idea that the work I did with Tim Bunch explains the large-scale structure of the universe as a quantum vacuum effect at the end of inflation, about a hundred trillion-trillion-trillionths of a second after the big bang origin. On the other hand, in the next chapter I will argue that quantum mechanics breaks down in complex systems. You can't get more complex than the universe, can you? But actually, I can have it both ways. The universe at the end of inflation was absurdly simple – just uniformly expanding almost completely empty space. That was when the quantum fluctuations were laid down. Complexity (in the form of galaxies, etc.) came later and served to congeal the primordial quantum fluctuations into the (classical) large-scale structure of the universe we see now. So, I think quantum cosmology – treating the whole universe as a quantum system – was fundamentally important for a split second at the beginning, but doesn't explain much after that. With one rather monumental exception: the quantum vacuum. You can't get much simpler than that.

The runaway universe

When I was a student, many astronomers thought the universe wouldn't go on expanding forever, but would reach a point of maximum distension, after which it would begin to contract, eventually falling back on itself to a big crunch, like the big bang in reverse, and at that point obliterate itself in a spacetime singularity. Today, few cosmologists suppose that the universe will in fact ever start to contract. Following observations announced in the late 1990s, it now seems that not only will the expansion continue forever, but the rate of expansion is actually speeding up. The entire cosmos seems to be in the grip of a titanic antigravity force, overwhelming the hefty gravitational pull of the 10^{50} tonnes of matter, the gravitating effect of which serves as a brake on the expansion. Physicists call the antigravity force 'dark energy', and its source is usually cited as one of the great unsolved problems of science. But you don't have to look very far to find a plausible culprit.

In the previous chapter, we saw how the quantum vacuum

is seething with energy, although its effects are woefully small over small distances, such as in the laboratory. But, accumulated across the whole cosmos, it could become significant. At first sight it looks like the quantum vacuum energy would only add to the gravity of the universe, not combat it. Energy is, after all, mass, according to Einstein's famous $E = mc^2$ equation. On this basis, the total vacuum energy in the universe ought to exert a strong gravitational *pull*, restraining the expansion still more. It shouldn't push. That simple argument fails, however, because in addition to the quantum vacuum possessing energy, it also has pressure, and in Einstein's general theory of relativity, pressure as well as energy is a source of gravitation. That fact – that pressure gravitates – usually surprises people, because when pressure is encountered in daily life through its mechanical action (for example, gas escaping a punctured tyre) it pushes outward. But it also attracts nearby objects via its gravitational effect. We don't notice the gravitating pull associated with pressure because it is extremely small in everyday life, but in the early universe, the pressure of the cosmic afterglow radiation was huge – the temperature was billions of degrees – and as a result, the gravitational pull of the afterglow's pressure outweighed even the gravity of all the densely packed matter. When it comes to the quantum vacuum, however, the pressure turns out to be *negative*, which means it *anti*gravitates, i.e. the quantum vacuum (negative) pressure gravitationally pushes rather than pulls. The quantum vacuum is therefore inherently self-repulsive: the expansion doesn't dilute the quantum vacuum energy but maintains it to be constant, creating ever more vacuum energy to fill it. And the arithmetic

is such that in the gigantic cosmic tussle of pushing and pulling, the push beats the pull, in a runaway process that looks
set to dominate the future of the universe.

Is the mysterious dark energy nothing more than the
quantum vacuum? It could be, but the case would be more
convincing if the numbers came out right. I explained on
p. 170 how the theory predicts that the amount of energy in the
quantum vacuum is technically limitless, leading physicists to
use a mathematical dodge and simply define whatever energy
is there when no *real* photons (or other particles) are present to be 'zero energy'. While it's true that ignoring an infinite amount of energy seems like playing fast and loose with
the laws of physics, it's harmless when only energy *differences*
matter. That was the case, for example, with the Casimir
force. However, when we want to take account of the gravitational effects of quantum vacuum energy, we can't be so cava-
lier, because *all* energy is supposed to gravitate. Does none
of that quantum vacuum energy gravitate? Maybe some of
it? How much? Since nobody knows the base level of quantum vacuum energy at which its gravitation effect kicks in,
the best one can do is just measure the rate at which the expansion of the universe is accelerating and work backwards
to calculate the energy density that would cause it. The astronomical observations imply that a cubic kilometre of space
contains about half a joule of dark energy. That's not much: it
would take a million cubic kilometres worth of dark energy to
boil a kettle. Yet integrated over the whole universe the total
adds up to more than all the other sources of mass-energy
put together, which is why dark energy now dominates the
dynamics of the universe.

What, then, is the ultimate fate of this incredible self-expanding space? If the quantum vacuum is indeed the primary explanation for dark energy, then the universe will eventually empty itself out of all matter and go on expanding for ever in a state of cold, dark, vacuum. If something else is at play, the universe may end abruptly with a big crunch (like the big bang in reverse) or a big rip, in which space expands so fast it comes apart at the seams and ceases to exist. Either way, the outlook is bleak, but it does raise some intriguing questions about the far future of the universe. What will happen to all the contents of the vast cosmos – the stars and galaxies, the gas and dust, and the atoms of which they are made? None of these things, it turns out, are for ever. But the objects with by far the greatest longevity are also the most enigmatic: black holes.

Black holes: what goes in must come out – or not

If the universe does go on expanding for ever, what will happen to all those black holes dotted across the cosmos? Their future, it turns out, depends critically on their mass. When the possibility of black holes was first discussed seriously in the 1960s, there was a lot of speculation about how they might form in the real universe. The favourite scenario was the implosion of a star, and indeed, the first black hole to be discovered (in the 1970s) was of this variety. It was also suggested that supermassive black holes might hide in the centres of galaxies: they too have been discovered in more recent years, although there is disagreement about how such monsters formed in the first place. But there was a third possibility. The big bang might have coughed out black holes formed in the first split second of cosmic history, from the gravitational implosion of ultra-dense pockets of material that filled the universe at that time. Theory suggested that these 'primordial' black holes could be very small – perhaps less than the size of an atom. Hawking did some of the early work on the topic, but then in 1974 found that mini-black holes have an in-built quantum self-destruct mechanism. They emit heat radiation (see p. 165), which causes them to shrink because they continuously lose energy, hence mass. The mass loss accelerates as a black hole gets smaller, terminating in an explosive burst of high energy particles leaving . . . what? Nothing at all? That, on the face of it, is what the theory says. But something about the incredible vanishing hole doesn't seem right.

Hawking derived his result by applying quantum mechanics, not to the black hole itself, but to the quantum vacuum that gets sucked into it. According to the rules of quantum mechanics, the wave function describing the electromagnetic field, for example, represents everything that can be known about it. And a wave function, like the proverbial elephant, never forgets. As it evolves with time, it keeps perfect tally of all physical variables. Trouble is, a black hole is by definition an object that swallows things and totally erases their identities. The surface of the black hole is called an event horizon precisely because nobody outside the hole can ever witness any events – anything at all – inside it. So the wave function that never forgets infuses the object that never remembers. Surely something has to give?

A black hole forms when stuff implodes – a star, for example – collapsing under gravity. Anything that falls in vanishes from the universe, although its mass remains discernible because the black hole gets a bit bigger with each sacrificed object. It's true that a stellar mass black hole, unlike the primordial mini-holes that Hawking first studied, will have an exceedingly low temperature. As a result, it would shrink at an unimaginably slow rate. But it will shrink. We can imagine a very, very far future in which there is no matter left to fall in, and the black hole just sits there, slowly emitting heat, getting smaller and smaller over the aeons, eventually evaporating completely. In summary, stuff goes in and stuff comes out. Remember: the wave function never forgets, so what comes out must include – maybe in a scrambled and barely recognizable form – what went in. Expressed more formally, the total information entering the black hole must equal the total information coming out in the Hawking radiation. At least, that is what quantum mechanics predicts. But Hawking's own 1974 calculation refutes that conclusion. Hawking's calculation implied that the emitted radiation has *zero* information about the past (technically, it is thermal radiation). Physicists refer to this mismatch as 'the black hole information paradox', which remains unresolved all these years later. Do black holes – extreme objects that they are – signal a breakdown of quantum mechanics? Does the infalling information vanish into nowhere, or does it sneak back out in a form so convoluted that the radiation *appears* random but in fact cryptically encodes information in complex correlations? Alternatively, is the black hole a portal to another universe where the information ends up? Or none of the above?

Personally, I think the black hole information paradox is an overblown fixation in the physics community. It is based on a highly idealized model in which the black hole is an isolated object and the quantum field is in a pure, completely defined state – which is of course totally unrealistic. In addition, the collapse of the star to form the black hole isn't included in the quantum analysis either. In short, the calculations leave out the rest of the universe. For quantum information to be *exactly* conserved for the black hole all possible external disturbances have to be ignored, which is a ludicrous oversimplification.

Even if the ultimate end state of a black hole remains a mystery, we can still be confident that it will indeed slowly shrink by the Hawking effect, over an immense time span. The numbers are mind-boggling. Even a tiny nuclear-sized primordial black hole will survive for billions of years. A solar mass black hole – common in the universe – would take 10^{67} years to evaporate and a billion solar mass black hole of the sort found at the centres of some galaxies would linger for an astonishing 10^{94} years.

It seems unlikely that we will get to the bottom of the black hole information paradox without a fully functional theory of quantum gravity, which so far eludes us. Straightforward attempts to apply quantum mechanics to the gravitational field by analogy with the electromagnetic field founder amid a plethora of infinite quantities and nonsensical results. Many alternative approaches have emerged – string theory and loop quantum gravity are two that attract a lot of attention. But they remain a work in progress. Nor can we look to experiment as a guide. Gravity is such a weak force,* any quantum

---

* Gravity is strong enough to keep our feet on the ground, but it takes a whole planet for us to feel it. The gravitational force between two electrons is a million trillion trillion trillion times smaller than their electric repulsion. That's weak.

effects are almost inconceivably small. For example, your kitchen light bulb *might* just emit a graviton – the quantum that does for gravity what the photon does for electromagnetism. How likely? Well, a rough-and-ready calculation I just did suggests that if humanity keeps the lights on at the present level for about 30,000 years, a single graviton might be created (and fly off into space unnoticed and uncelebrated). It is frustrating that after decades of effort, two of the greatest advances of twentieth-century physics, quantum mechanics and Einstein's general relativity theory of gravitation, haven't been unified. Is that because there is something wrong with our understanding of gravitation? Or perhaps something wrong with quantum mechanics?

It's time to take stock.

PART FOUR

Philosophical Finale: The Search for Meaning

CHAPTER 10

Weirder Still

Quantum mechanics is without doubt the most successful theory in history. It has been tested again and again in rigorous laboratory settings and through myriad daily applications of quantum technology. It has never been known to fail. In science, however, no theory should be so entrenched that it need not be called into question, no experiment so exacting that further scrutiny is unnecessary. In recent years, new and more powerful tests have been devised to investigate overlooked or subtle aspects of quantum mechanics and to check its predictions in the most extreme attainable circumstances. These experiments have revealed that the quantum world is even weirder than we thought. And if there are any cracks in the theory, they have yet to become manifest. But in spite of that, further tests are being devised. There has always been a minority who think quantum mechanics is simply too crazy to be right. After all, here is a theory that, at rock bottom, defies any straightforward understanding in normal terms and can sometimes appear out-and-out nonsensical: particles in many places at once, spooky telepathic links across space, boisterous atoms that freeze when they're being watched . . . What to make of it?

Boiled down to its essentials, the problem is this. Quantum mechanics describes the microworld with spectacular success in terms of *blended realities*. Yet we uncover only a single concrete reality when we probe the murky quantum domain. Something projects out one actual 'real world' from a ghostly amalgam of many possibilities. But how? What are these 'parallel realities' and can they affect anything we can directly measure? Addressing these questions takes us into some very strange territory.

The dog that didn't bark

In the Sherlock Holmes story 'Silver Blaze', the famous detective uses his formidable deductive skills to solve the mystery of a missing racehorse and a murdered trainer. The vital clue was provided by the dog that didn't bark in the night. In Chapter 2, I described the watched pot effect: how a detector that fails to register a photon from an excited atom indicates that the atom has not decayed, and so re-sets the atom into its initial excited state (see p. 38). A quantum *non*-event therefore serves as an actual measurement – that the atom is still excited – and thereby 'collapses the wave function' back to one describing the initial excited state. In the extreme case of continuous measurement – never-ending non-events – it paralyses the poor atom in a state of permanent excitation. But how can something that *isn't there* make a difference to material objects?

In philosophy, 'counterfactuals' are hypothetical statements that explore what would have happened if certain events or conditions had been different from what actually occurred. We use counterfactual reasoning all the time ('If only I hadn't missed the train, I'd be married to Ann instead of Betty . . .'). Being able to foresee the consequences of alternative scenarios influences a great deal of decision-making, moral reasoning and historical analysis. But it also plays a critical role in quantum mechanics, where non-events can have actual consequences for material objects, simply as a result of something that might have been, but wasn't.

Let me explain how this comes about by describing a striking demonstration: an experiment by Anton Zeilinger and his colleagues, then at the University of Innsbruck in Austria. Their apparatus is shown schematically in Fig. 15, which features an upper channel and a lower channel along which light may pass. A single photon enters from bottom left and encounters a beam-splitter that transmits with only a small probability (for example, using a type of mirror with 95 per cent reflectivity). Chances are the photon won't enter the upper channel, and if it doesn't, it gets reflected back down and then up again to a second beam-splitter for another attempt, and then a third and so on. After sufficiently many tries, it's very likely that the photon will finally have made it to the upper channel and will exit to the top right to be captured by an awaiting detector, as illustrated in 15(a). However, if you decide to take peeks at the photon's progress by stationing multiple photon detectors along the upper channel (Fig. 15(b)), a non-detection after each attempt the photon makes to get upstairs will ensure that the photon stays downstairs (by collapsing the wave function, watched-pot-like, to a downstairs state and re-setting). As a result, the photon will exit *bottom* right. It seems that the mere *threat* of detection upstairs is enough to keep the photon cowering in the lower channel.[1]

The Quantum Measurement Laboratory at Imperial College London is dedicated to such quantum peculiarities with a view to practical applications. The lab's Michael Vanner

(a)

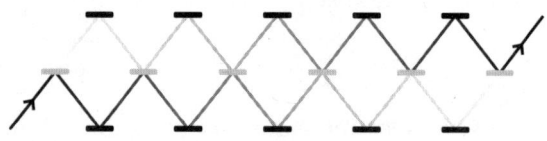

(b)

Photon detectors

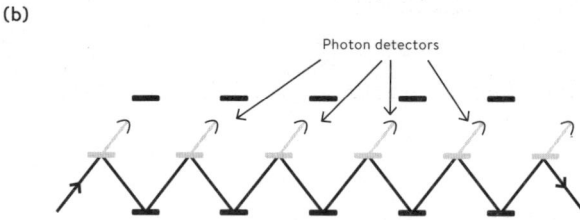

Figure 15

A photon enters the track from the bottom left and hits a reflector which has a very small transparency, meaning the photon *might* pass through into the upper channel, but with low probability. If it doesn't, it gets reflected back for another try, and another . . . After many tries, the photon is very likely to have penetrated to the upper channel, where it will be detected exiting from the system at top right. In the modification (b) a series of detectors (shown as a row of little inverted cups) is placed to inquire about the success or failure of each 'try'. Each *non*-detection of a penetrating photon serves to re-set the wave function to the lower channel and effectively trap the photon on the lower level, where it exits bottom right. The experiment shows how registering the *absence* of a photon can have physical consequences in another part of the system.

reported an experiment that exploits the non-detection of photons to produce a type of quantum refrigeration process,[2] by stationing a photon detector near a tiny glass bead and registering the absence of photons. The resulting Zeno effect measurement paralysis calmed the internal vibrations of the bead beyond the capability of other methods. 'This work utilizes the counter-intuitive fact that a measurement of "nothing" can have a significant impact to the state of a physical system,' Vanner and his colleagues point out. They regard quantum measurement as 'a powerful resource . . . for quantum-state engineering and quantum information applications,' such as information storage in a quantum computer.

To highlight the profound oddity of counterfactuals that make a physical difference, or, how what might-have-been can be as important as what-actually-was, the physicists Avshalom Elitzur and Lev Vaidman of Tel Aviv University dreamed up a hypothetical scenario wherein the *failure* of a photon to be detected by a target could nevertheless be used to test whether a bomb would have exploded if the photon *had* been detected. On the scale of quantum weirdness, this is one that surely counts as a front-runner.[3]

The key to quantum counterfactuals is the crucial property of superposition. Remember, in quantum mechanics a single particle like a photon might take this path or it might take that path, but one may also engineer a superposition in which the photon can (in some sense) take both paths. A better way to think about it is that each path is taken by a 'ghost' of the photon and the two ghosts can later be recombined. But – this is paramount – the experiment works only if nobody peeks to see which path the photon *actually* takes, because that would serve to collapse the wave function and instantly exorcise the other path's ghost. Splitting and recombining light beams is a standard procedure in physics labs and a lot of off-the-shelf commercial devices, called interferometers, utilize it. (I described a gargantuan interferometer called LIGO in Chapter 6.) The Elitzur–Vaidman bomb-testing device is an adaptation of a type of interferometer that demonstrates the full magic of the quantum world. The details are a bit technical, so the reader may wish to skip to the next section, but the conclusion is so shocking it's worth persevering.

Like the proverbial magic show, the bomb-testing trick is all done with mirrors. The arrangement is depicted in Fig. 16. First look at (a): a single photon entering from bottom left hits a beam-splitter A which lets half the light through and deflects the other half to the left. When just one photon is involved,

Figure 16

Using ghost photons to test for working bombs. A photon enters from bottom left and hits a beam-splitter, A. The photon is now in a superposition state with one 'ghost' (i.e. component of the superposition) taking the upper path and the other the lower path. In Fig 16(b) the ghosts are reunited at the second beam-splitter, B. The phases of the waves are such that the ghosts reinforce and the photon will go 100 per cent of the time to photon detector D_1. However, if an obstacle like a brick is placed in the upper path as in 16(c), it kills the upper-path ghost (Ghost 2). The other, lower-path, 'bereaved' ghost will then be on its own, unencumbered and free to choose. When it encounters B, it will go to D_1 about half the time and straight ahead to D_2 the other times. If the brick is now replaced by a photosensitive bomb, about half the time the bomb explodes, but a quarter of the time a photon is detected at D_2 even though the bomb did not explode. The detected photon registered the presence of a working bomb, even though the photon failed to detonate it.

(a)

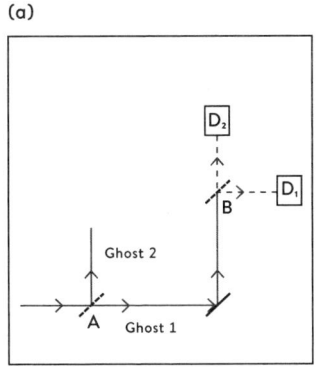

(b)

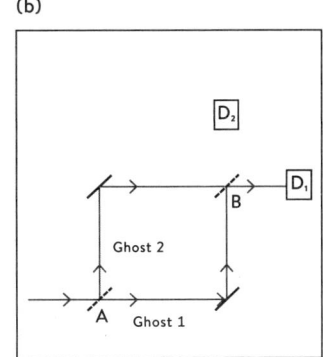

(c)

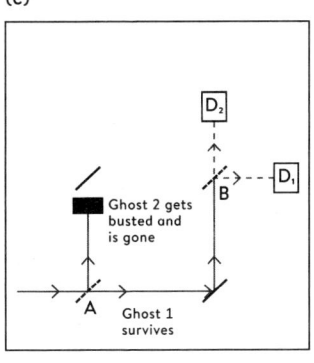

(d)

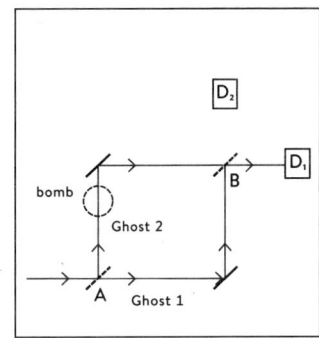

the splitter ensures that the photon is transmitted or reflected with equal probability, should anybody perform a detection measurement to find out. If nobody does, the photon 'dematerializes' into a superposition of two 'ghosts', one going straight through A and the other being deflected left (vertically in the figure as shown). I've labelled them Ghost 1 and Ghost 2 respectively. A moment later, Ghost 1 is also redirected left (up in the figure) by a regular 100 per cent reflecting mirror, where it encounters a second beam-splitter, B. There Ghost 1 splits again, going straight with 50 per cent probability, or to the right with 50 per cent probability. Photon detectors, labelled D_1 and D_2, are stationed to find out what happens, and each has a ¼ probability of detecting a photon, because Ghost 1 is itself only one-half of the original photon superposition; the other half is lost in the set-up shown in Fig. 16(a).

Next look at Fig. 16(b), in which Ghost 2 now joins the action. A second regular mirror has been added (at top left) to redirect Ghost 2 to the beam-splitter B from a different direction, where it too will be partly transmitted and partly reflected to the left. At this juncture, the two ghosts meet up and merge. This is where the wave nature of the photon is critical. Each ghost may be described by a wave, and if the distances are arranged correctly, the transmitted (by beam-splitter B) component of Ghost wave 1 and the reflected component of Ghost wave 2 (i.e. those parts of each heading towards detector D_2) can be deliberately placed out of phase with each other, so that the peaks of one ghost wave will be aligned with the troughs of the other. The result is that the wave is cancelled – recall this is called destructive interference (p. 43). So D_2 remains silent: no wave reaches it, so no photon triggers it. But the ghost waves heading to D_1 have the converse property: the reflected part of Ghost wave 1 and the transmitted part of Ghost wave 2 arrive peak to peak, trough to trough, so they reinforce (constructive interference). Therefore, D_1 clicks, registering the arrival of the photon. It has gone from bottom left to top right in the figure, by two separate paths, split into a ghostly superposition and then reconstituted into a single, whole, real photon.

So far, I have described a fairly conventional (and useful) piece of optical apparatus; there's nothing truly fishy about it. But there can be a very odd twist. The failure of D_2 to detect a photon hinges on both paths of the interferometer remaining open. If something blocks the path of Ghost 2, preventing it from reaching B, there will no longer be interference and the situation will revert to that of Fig. 16(a) where both D_1 and D_2 might click, with equal probability in the 50 per cent of cases that the photon isn't absorbed en route. Figure 16(c) depicts that scenario, where an object (such as a brick) blocks the path of Ghost 2, absorbing and thereby annihilating it. If the experimenter notices D_2 has clicked (which is expected to happen ¼ of the time on average) they could deduce that someone had obstructed one of the beams, without needing to actually look for a brick. We now reach the punchline of this protracted description: as far as D_2 is concerned, no brick, no click. Ever. If D_2 does click, it tells you that a brick (or something) is blocking one of the paths, but at the same time it also confirms that the said photon has survived the journey and arrived safely at the detector: it cannot have been absorbed by a brick en route. And yet the presence of the brick still affects the result – by permitting a click in D_2 – even though just a single photon enters the apparatus and a single photon exits it. Ghost 1 evidently senses the brick's presence in the other path (the road not travelled) in a type of quantum clairvoyance, or 'sniffing out at a distance'.

How can that be? What is going on here? To elucidate it, Elitzur and Vaidman envisaged replacing the brick with a bomb; see Fig. 16(d). Imagine a manufacturer that makes a light-activated bomb triggered by a super-sensitive photon-detector (similar to D_1 or D_2) perched on top, wired to the detonator. You'd need to keep such a bomb in a light-tight box, or it would explode. Suppose the trigger was so sensitive that even a single photon hitting the bomb's detector was enough to set the bomb off. In this story, the manufacturer has delivered 100 bombs to the army, but due to a flaw in the manufacturing process, it is feared that some of the bombs may be defective because the attached photon detector is non-functional. The conundrum is, which? If you open the boxes one by one to look, any good bomb will immediately explode. But using the interferometer in Fig. 16, there is a way to identify good bombs without exploding them, which seems paradoxical.

This is what you do. In pitch darkness, you insert the bomb into the interferometer, aligned so that the bomb's photon-detector will intercept Ghost 2. What happens? If the sensor on the bomb is defective and just lets the light go on through, nothing happens – it's as if the bomb wasn't there, and D_1 clicks to register the photon's arrival in the usual way. But suppose the bomb mechanism is functional? From the statistics as I described them above, about half the time the functional bomb will explode because Ghost 2 triggered it. Half the time? Yes, remember Ghost 2 is just a probability wave, not a real entity in the everyday sense; it has the potential to trigger the mechanism, but also the potential to not trigger it, in equal measure. (Triggering the bomb has a rather drastic consequence for the lab staff, who will need to take some protective measures.) The good news is that there is a roughly 50 per cent chance that the functional bomb will not explode; Ghost 2 passes by, meets the working detector, but in that great cosmic lottery called quantum uncertainty, fails to be 'detected', and so continues on its way to mirror B. But, as I have explained, the *failure* to explode the bomb counts as a measurement – which has consequences. The downstream form of the Ghost 2 wave is altered, disrupting the delicate phase-cancelling properties between Ghost 1 and Ghost 2 when they are recombined. As a result, the wave phases of Ghost 1 and the now-modified Ghost 2 will combine differently at the mirror B. Instead of always interfering constructively in the direction of D_1 and destructively in the direction of D_2, they combine to yield equal strength waves in *both* directions, i.e. towards both detectors. As a result, there is a 50 per cent chance that D_2 will click, indicating 'working bomb encountered by Ghost 2'. Unfortunately, there is also a 50 per cent chance that D_1 will click, which is no help, as you can't tell from that result whether the bomb is indeed functional and detected by Ghost 2 or is simply a defective one. So you would need to keep repeating the experiment with that bomb in the case that D_1 clicks, until you were pretty sure about it one way or the other.

Whew!

Here is a summary of the precise statistics for the functional bombs. The (average) total apportionment per trial is: one-half of the time, no clicks, bomb explodes; one quarter of the time, D_2 clicks, and a working bomb is detected; one quarter of the time, D_1 clicks, inconclusive result, repeat test. The upshot is that you have used a photon to verify that a photon-triggered bomb works without in fact exploding it! The bad bombs are thereby identified and discarded and a fraction of the good bombs can be retained for use. (The rest have blown up.) What this boils down to is that detector D_2 clicks because the bomb could have exploded but didn't. It describes a sort of parallel reality that bleeds into this reality and actually makes a physical difference. A non-barking quantum dog bomb.

Elitzur and colleagues put it this way:

'A promising candidate for a "master quantum oddity" is the unique status of the non-event, i.e. a "counterfactual". This is an event which could have happened but did not. Quantum non-events are unique in that they can exert significant causal effects just by virtue of this "could have".'

Backwards in time? The ghosts of photons past

If non-events can make a physical difference, what does that imply about past events? Did they 'really happen', and how would we know? The essence of quantum mechanics is uncertainty. Heisenberg's uncertainty principle quantifies the expanding cone of ignorance going forward in time. But it works in both directions: the cone spreads out backwards in time too. In quantum mechanics, the past is generally not a done deal; it, too, is uncertain. As I described in Chapter 2, an electron is *here* now, but where was it a minute ago? Quantum mechanics says there may be no answer. Or again, it behaves like a particle when a certain measurement is made; but was it a particle or a wave one second ago? No answer. It seems that there is no fixed history connecting the past to the present, only multiple possible 'ghost' histories in a gigantic superposition.

In 1978, John Wheeler went so far as to claim that 'the past has no existence except as it is recorded in the present'.[4] He arrived at this arresting conclusion by analysing a variant of the two-slit experiment (recall Fig. 7). To recap, if both slits are left open, the photons hitting the image screen make a speckled pattern of bright and dark bands, or fringes, in accordance with the wavelike aspect of light; each photon has two ghosts, one of which goes through one slit and the other through the other, then they recombine at the image screen, and collectively, after many photons have traversed the system, they produce an interference pattern. When one slit is closed, the interference pattern goes away. If the experimenter stations detectors near the slits so as to determine to which slit any given photon is headed, one by one, no interference occurs when the results are aggregated. Obtaining definite knowledge of the photon's path collapses the wave function and exorcises the other ghost. As a result, the photons behave like particles. All this is established wisdom.

In the standard set-up, the experimenter decides beforehand whether to position tell-tale detectors or not, and thus which aspect of reality – wave or particle – shall thereby be manifested. But Wheeler pointed out that a decision on which type of experimental arrangement to use could actually be made *after* the light had traversed the screen with the slits, at a time when one might have imagined that the reality die (wave or particle) was already cast. In a cartoonish description of how this is done, imagine that the image screen is replaced by a pair of telescopes, each trained on one slit, so the experimenter could look back and determine through which slit any given photon had come (see Fig. 17). That would reveal the particle-like form of light. But if, at the last moment, the experimenter inserted a screen to block the backward view, then the arriving photons dappling the screen would build up a wavelike interference pattern. The experimenter could flip-flop back and forth – any photons that arrived with the screen in place would be part of a wave description of reality, any that entered one or other of the telescopes would be part of a particle description.

Now, you might think that any given photon would 'know' whether it went through one slit or both; a single photon 'going through both slits' means that one ghost goes through one slit and the other ghost goes through the other. The startling feature of the delayed-choice experiment is that the disclosed reality – wave or particle – is determined only *after* the light has passed through the slits. That is to say, if the experimenter chooses to look back through a telescope and sees the photon coming from slit A, then that act of observation determines the nature of the photon's reality at a time *prior* to the observation. So what an experimenter does now affects the nature of reality that *was* – in the experimenter's past. Evidently, the act of observation that snaps quantum fuzziness into definite reality can reach back in time. It's true that in a typical Young's experiment this 'reach-back' duration is mere nanoseconds, but still, in principle the reach-back time could be as long as you like.

It could in principle be the age of the universe, as I first learned from Wheeler himself, over breakfast at a physics

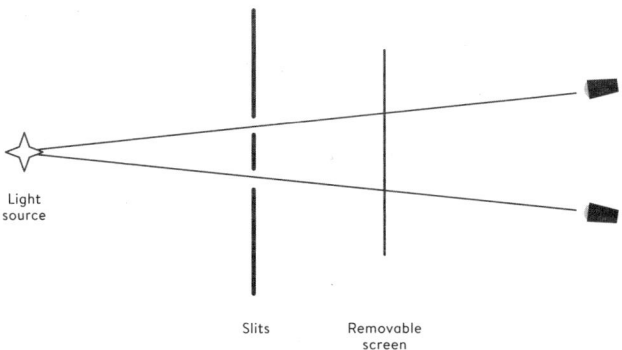

Light
source

Slits

Removable
screen

Figure 17

Delayed-choice experiment: the past is not a done deal! In this variant of the two-slit system, the observer has the choice of looking back through the telescopes to determine which slit a photon has passed through, or inserting a screen that blocks the view. In the latter case, the photons cumulatively make an interference pattern on the screen, indicating that they are behaving like waves, but in the former case they do not, indicating that those photons behave like particles. Wave-particle duality – you the experimenter can choose which aspect to observe. But – and this is the key refinement – the choice of wave or particle is not made until *after* the photons have traversed the slit screen and are well on their way to the detectors. Your later decision determines whether a given photon *shall have been* a wave or a particle in the (albeit recent) past. The nature of past reality is influenced by future decisions. However, the experiment cannot be used to change the past or to send information backwards in time.

conference in Baltimore. 'How do you hold up the ghost of a photon?' he asked me, cryptically, as I poured my coffee. The question was typical of Wheeler's playful, pithy style. It took me a while to twig that he was envisaging a Young's two-slit experiment over astronomical distances, with the gravitation of a distant galaxy bending the photon's path in place of the slits in a screen. That would likely result in an arrival time difference at Earth between the two ghosts of about a month – hence the need to 'hold up' one of them until the other came along to merge with the first in an interference experiment, revealing the photon's wavelike nature. Or – it could be Wheeler's choice – *not* merge the ghosts and obtain a particle path (either one side of the galaxy or the other). The choice in this case would have a 'reality reach-back time' of billions of years!

I left the discussion thinking that the delayed-choice variant of Young's experiment was just a nit-picky detail in an already exhaustively analysed experiment, unlikely ever to be performed or to yield anything significant. I was wrong. Within a few years, it became possible to test Wheeler's thought experiment, albeit in a laboratory and not on a cosmic scale. Rather than the Young's two-slit apparatus, a better method is to use something like the 'bomb-testing' interferometer described in the previous section. With such an arrangement, the delayed choice to be made by the experimenter is whether or not to insert the second beam-splitter B in Fig. 16(b). Without B, there is no interference between the two ghosts, and the photons go about half the time to D_1, and half to D_2, in accordance with them behaving like particles that take *either* the lower path or the upper path in the figure, respectively. But recall that when B is inserted, each photon behaves like a wave, their ghost parts taking *both* the upper and lower routes, and interfering when they merge at B. As a result, they go 100 per cent of the time to D_1. Crucially, the experimenter does not need to make the decision 'I will/won't insert B' until *after* the given photon has entered the interferometer and is well on its way towards the detectors.

In the above description, I have given the impression of an eager physicist clutching a beam-splitter, dithering over whether or not to stick it into the interferometer at the last moment, which in principle is certainly a possible procedure, were things not happening so fast. In practice, the whole process is high-speed automated and randomized, because the photons arrive within nanoseconds. Anyway, the experiments confirm[5] that random 'last-minute' changes (B in or out) do affect whether any given photon *shall have* taken both paths or one. Wheeler was right!

Erasing the past

As if the foregoing results are not bewildering enough, there's a further mind-bending twist. You can cook up an experiment whereby the 'which path' information is first garnered, but then erased. The experimenter gets only a brief window of opportunity to peek at the result before it vanishes. The choice this time is whether the experimenter decides to look, thus observing a particle-like photon coming from an identifiable slit, or to pass up that opportunity and deduce that the photon has behaved like a wave going through both slits. In 1982, a quantum-eraser, delayed-choice experiment was proposed by Marlan Scully and his colleagues at Texas A & M University[6] using an adaptation of the interferometer I described above.

In this Scully experiment, a 'proxy photon' goes off somewhere while its twin goes through the slit system (see p. 49). To make the results more graphic, imagine two observers, Alice and Bob, stationed as shown in Fig. 18, with Alice able to detect the slit-system photon and Bob the proxy photon (labelled 'Idler' in the picture). With such a set-up, Bob can decide to inspect the proxy photon or not. Alice has a detector D_o, which is moveable, so she can scan across the line of sight and record the spread in the pattern of arrivals to look for interference fringes in the accumulated results. She won't see such a pattern if Bob inspects his idler photon. But Bob may instead decide to *not* inspect his photon, in which case, Alice *will* observe an interference pattern (again, after many such events are combined). Bob doesn't have to decide in each case until Alice has *already detected* and recorded her photon. To take it to the extreme, if the photons heading Bob's way could be retained safely (i.e. coherently), Bob doesn't have to act right away; he could wait until after his lunch break to make up his mind about inspecting them or not on a one-by-one basis.

In the real experiment (which was done at the University of Maryland without lunch breaks) an automated system toggled randomly between the detectors, registering the proxy photon or not.[7] There are then two sets of results randomly intermingled – those where the 'which-slit' information is recorded, and those where it isn't. Sure enough, when the results are untangled, there is an interference pattern in one set of data but not in the other. In case you are wondering how the 'will-look, won't-look' alternatives are arranged in the actual experiment, it is by using a prism to make the paths of the photons coming from one slit or the other diverge so

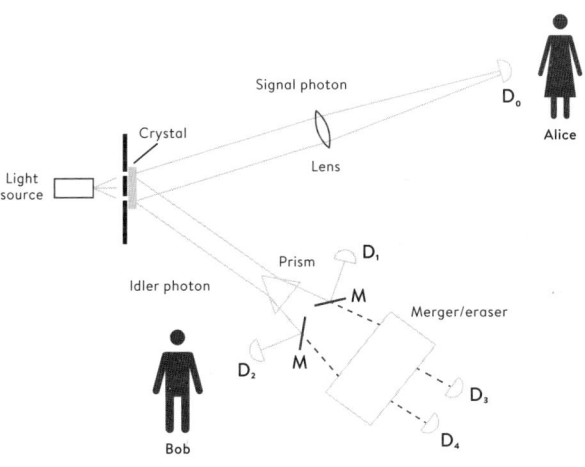

Figure 18

Quantum procrastination. In this 'quantum eraser' experiment, each photon passing through a two-slit experiment is converted by a special crystal into two entangled photons, one of which – the signal photon – goes to detector D_0, which cannot distinguish which slit that photon passed through. The other – the idler photon – goes through a prism that separates the two possible paths so that a pair of detectors (one for each path) *can* tell which slit the original photon had earlier traversed. However, this path discrimination is disclosed only by flicking a (metaphorical) switch to direct the idler photons with mirrors M to detectors D_1 or D_2. In absence of the flicking, the two idler paths get conflated inside the black box, and as a result the 'which-slit' information gets permanently erased. The experimenter can dither and make the choice 'to see or not to see' until after the original photon has passed through the slit screen.

the difference is detectable, but then re-merging the paths in the case of 'won't look'. And actually, it's really a case of can't-look-won't-look (to paraphrase the title of Dario Fo's famous play), because the merging of the proxy photon paths irretrievably erases the which-slit information. So the which-slit result is potentially available for a while, and can be inspected if the experimenter is quick off the mark, but then the information gets permanently obliterated.

If you feel baffled by this account, you are not alone; even Alice and Bob are flummoxed. So for readers who are determined to get to the bottom of the whole saga, let me try to unpack the details. The actual experiment, performed in 1999 by Yoon-Ho Kim and colleagues, is illustrated in Fig. 18, which I have simplified to omit some technicalities.[8] A photon passes through the two-slit screen and encounters a crystal that turns one photon into an entangled pair of progeny photons that fly out in different directions. Because they are entangled, inspecting one tells you something about the other. Alice is charged with detecting the incoming signal photons one by one using the moveable detector D_o. The merging lens stationed between Alice and the slits ensures that her equipment alone cannot be used to discriminate from which slit any given photon emanates. Bob, however, *can* tell which slit each photon went through because at his end of the lab, in place of the converging lens, there is a diverging prism that has the effect of spreading out the two possible paths. He can use mirrors M to direct his photon to detector D_1 in event that the photon comes from the upper slit (as shown in Fig. 18), or D_2 if it comes from the lower slit. In this configuration, Alice shouldn't detect any interference pattern, because Bob has found out the proxy photon's path and, because of the entanglement, thereby also knows which slit the signal photon went though, even though Alice doesn't. Sure enough, when Alice collects all her detector data, she gets a graph like Fig. 19, showing two slightly overlapping humps (representing two side-by-side bands of light), and no sign of interference fringes of the sort apparent in Fig. 7.

So far, so expected. But Bob has a choice. (This is a delayed-choice experiment, remember.) He can whisk away the mirrors M and thus relinquish the opportunity to obtain which-slit information: his detectors D_1 and D_2 remain silent. In that event, the photon can continue on its merry way. Soon, it encounters an optical system that I've put in a box to avoid complicating the picture. The box contains 'the eraser'. This is not a photon annihilator, like the brick in Fig. 16(c), but a cunning device to mix and match the two separated (by the prism) photon paths so that they can no longer be distinguished. It is the path information that is erased, not the actual photon, which emerges to be detected at either D_3 or D_4. But the optical design is such that a photon from either slit can hit either detector D_3 or D_4 with equal probability. If Bob makes this choice (to remove mirrors M) then he can no longer discover which slit any given photon emanated from, and so he concludes that Alice *should* see interference fringes for those particular paired photons. And since Bob need not make up his mind about the fate of any given idler photon until *after* Alice has detected and recorded her signal photon, it seems like Bob is either wrong in his prediction (using quantum mechanics), or that his actions have somehow reached back in time to determine Alice's results. Seth Lloyd has coined the term 'quantum procrastination' for these investigations in which Bob dithers,

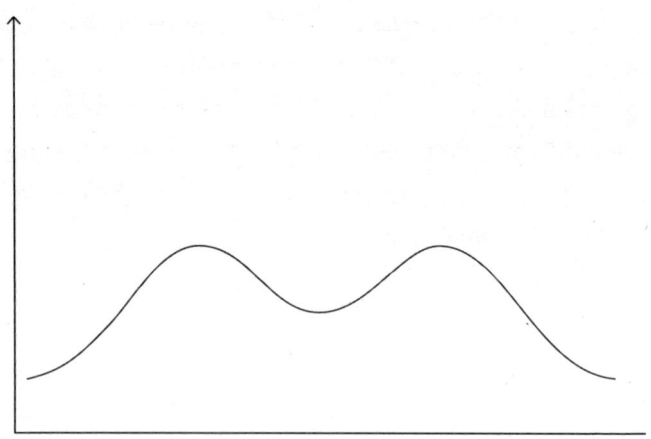

Figure 19
The twin peaks show the graph that Alice obtains by
counting the accumulating signal photons that come her
way, scanning across the line of sight to cover the spread.
The peaks graphically represent a pair of bright bands,
one from each slit, that slightly overlap, but do not show
an interference pattern that would indicate wavelike
behaviour.

pointing out that: 'It is the measurement on the second [distant] photon – apparently retroactively – that made interference take place or not.'[9]

Perhaps disappointingly, there is in fact no actual retro-causation going on here. What Alice got, Alice got. Her results (two humps) remain unchanged whatever Bob decides to do later. What, then, of the expected interference fringes? Where are they? It turns out they are there alright, but need to be teased out of Alice's data. Photons don't come with a label 'Beware, I am entangled!'. Alice can only tell which of her photons is entangled with what and where by sitting down with Bob (after lunch) and comparing notes. That way they can marry up signal and idler photons pair by pair and sort them into groups. One group includes the results when Bob chose to use the mirrors. No mystery there; Bob predicts no interference fringes, just two humps like those that Alice in fact measured, because Bob knew which slit each of those photons traversed and was thus able to observe the particle-like nature of each. But how about the cases where Bob *didn't* use the mirrors, and let the photons fly on through the merger/eraser black box to hit D_3 or D_4? Merging both sets of results from those two detectors and cross-referencing with Alice yields nothing new, just the two boring humps again. However, if Alice extracts only those photons she detected at D_0 that correlate with the photons Bob detected at D_3 alone, an amazing result leaps out. She finds that that subset of photons forms a pattern like the one shown in Fig. 20, curve A. Interference fringes, there for all to see! What about the data from D_4? Same thing, but slightly displaced (broken line, curve B); that is, out of phase with A. And when Alice amalgamates those two

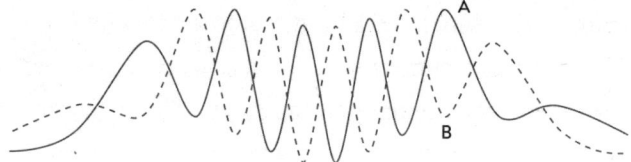

Figure 20
When Alice extracts from her data set only the photons that
correlate with Bob's detector D_3, an interference pattern emerges
in the graph (curve A). A similar pattern is obtained by matching
Alice's photon detections with Bob's results at D_4 (curve B). The two
curves are out of phase, and when combined they re-create the
double-hump patterns shown in Fig. 19. Two complementary wavelike
manifestations of light are buried in purely particle-like data!

patterns, she gets back the twin-humps pattern shown in Fig. 19. Who would have guessed that embedded in the humps indicating particle-like behaviour are two complementary patterns indicating wavelike behaviour?*

A final comment: in the real experiment, there is no Alice and no procrastinating Bob meddling with mirrors, only lots of fancy automated optical apparatus. The correlations are logged using coincidence counters that pick out only the D_0 photons that match up either with D_1 or D_2 detection events or, suitably delayed, D_3 or D_4 events.

The fact that the interference pattern can still be restored even after the particle has ostensibly 'decided' which path to take certainly challenges our intuition about time and causality. It shows that quantum events are not always constrained by a straightforward temporal order; instead, the choice of measurement can seem to retroactively influence (though not alter) past events, blurring the distinction between cause and effect. Although the delayed-choice and quantum-eraser experiments were performed to test the foundations and consistency of quantum mechanics – especially the concepts of entanglement and superposition – they do form the basis for what has been described as 'Stephen Hawking's final theory'.[10] Hawking, in collaboration with Thomas Hertog, applied delayed-choice ideas to cosmology in an attempt to explain why the universe seems suspiciously rigged in favour of life (a much discussed and contentious topic that goes by the

---

* Could Bob not have deduced from which detector, D_3 or D_4, any given idler photon hit, which slit it went through? No. The output from the black box – which of D_3 or D_4 a photon will hit – depends on the wave phase of the photon, and the phase information cannot on its own reveal the which-slit information.

poor name of the Anthropic Principle). Only a life-friendly universe would spawn observers that could do the quantum-resolving bringing-into-being of the cosmic reality. The sticking point in explaining this happy circumstance was finding a way to use quantum measurements to select, not just for states of matter (like a particle-at-a-place), but for the very life-encouraging laws of physics themselves. That is what Hawking and Hertog attempted in what they called a 'top-down' theory of cosmology. Many years ago, I outlined similar ideas myself in my book *The Goldilocks Enigma*.[11]

Sneak peek: lifting the veil on quantum uncertainty

Given the pivotal role that measurement plays in projecting out concrete reality from quantum haziness, you might be wondering what, precisely, constitutes 'a measurement'? In practice, a quantum-measuring device is usually a complicated piece of macroscopic equipment. But whatever apparatus is used, the basic principle is the same: you have to couple the equipment to the quantum system in a way that collapses the wave function and determines a value for the quantity you wish to measure. The coupled system evolves so that something like a pointer or read-out on the apparatus becomes correlated with the value that the quantum system snaps into. For example, atom excited, pointer on the left, atom decayed, pointer on the right. The essence of a quantum measurement is that tiny quantum effects get hugely (and irreversibly) amplified to everyday size where the result can readily be inspected by the experimenter. (And communicated to a colleague in plain language, as Bohr put it.)

Buried in the foregoing description of a measurement is a crucial word: 'coupled'. What the apparatus does must be sensitive to what the quantum system does. The tiny quantum object has to be able to trigger a noticeable change in a big chunk of stuff. If the apparatus isn't in some way connected to the quantum system, no measurement can be made. But the vital coupling isn't an all-or-nothing affair. The linkage could be made very weak. In that case, the disturbance – sometimes called the back-reaction – occasioned by the measurement may not completely collapse the wave function. Indeed, for exceedingly weak coupling, the quantum state may not be disturbed significantly at all. This is called a *weak measurement*. Doesn't this mean that we can, after all, find out what a quantum system is doing without irreversibly changing it, by using a measuring device to sense the state of an atom, but with its crucial sensor turned so far down that it's almost switched off? And if we can do that and still get a result, does that not imply that the quantum system must, after all, possess actual values for properties like position and momentum all along, values that we could (with patience) glimpse using weak measurements only, avoiding full-blooded state-wrecking wave function-collapsing strong measurements? Weak measurements, it seems, could lift the veil of uncertainty that shrouds quantum systems.

There is, inevitably, a snag. If the coupling between the apparatus and the quantum system is weak, the measurement loses precision. The upshot is that the results will be scattered statistically over a range of values. In the extreme case of zero coupling, the measurement will yield nothing of significance to the experimenter at all. However, imprecise measurements are a feature of all science, and there is a straightforward way of tackling it: you repeat the measurements many times and use statistics to obtain a sharper result.

A famous example concerns Sir Francis Galton (Charles Darwin's half-cousin) and 787 members of the 1907 West of England Fat Stock and Poultry Society country fair in Plymouth. Galton invited the local townsfolk to guess the weight of an ox. A prize went to the person with the nearest guess. Galton was astonished to discover that although nobody got exactly the right answer, the *average* of all the guesses was very close to the correct weight of the ox. This property of averaging many guesses to get the truth is sometimes called the wisdom of crowds. And the greater the number of rough-and-ready statistically scattered measurements, the closer the average gets to the true value. Well, in the same vein, one can average many weak measurements of a quantum system. Would that not yield something like 'the true value' of, say, the energy of an atom, without disturbing it and collapsing its wave function?

The answer is yes and no. In the case of the ox, many people can estimate its weight simultaneously. That's impossible to do with an atom, but you can certainly prepare a large number of atoms in identical states and weakly measure the energy of each of them separately, then combine the results. If you do that, the answer is generally *not* one of the allowed energy levels. It will be some 'crazy' value between, or outside the range of, the permitted energies. For example, if the atom is in an equal superposition of excited state 1 and excited state 2, then a strong measurement will find the atom to be either precisely in state 1 with a probability of 0.5 or precisely in state 2 with a probability of 0.5. But a series of weak measurements will, when averaged, yield a value part way between the energies of states 1 and 2. It would certainly be wrong to conclude that an atom in the said superposition state has an electron in an orbit hovering between the allowed energy levels 1 and 2. In fact, it is impossible to conclude anything about an *individual* atom based on the average of weak measurements over an ensemble of many atoms. In spite of this shortcoming, weak measurements do yield meaningful additional information about quantum systems. Many such weak measurements have been performed experimentally and compared with theory. They do not contradict quantum mechanics, but open up a new window to help us frame a fuller picture of the quantum world.

Cheshire cats and other out-of-body experiences

A familiar feature of London is the iconic red bus. Occasionally you see them in other places; for example, I recently saw one in Guatemala City. It is a truism that where the bus goes, the red goes. It would make no sense to say the bus went to Guatemala but the red stayed in London. Yet that is exactly what quantum weak measurements suggest: particles can be spatially separated from the properties that define them. And experiments have confirmed this.

In Lewis Carroll's tale *Alice's Adventures in Wonderland*, Alice meets the Cheshire cat which slowly fades away, leaving only the grin. The cat is gone but the grin remains. Alice exclaims: 'Well! I've often seen a cat without a grin, but a grin without a cat! It's the most curious thing I ever saw in my life!'

By analogy, quantum experiments that separate particles from their properties are referred to as Cheshire cat states. One such state to be created succeeded in separating a neutron from its magnetic field.[12] (Although neutrons are electrically neutral overall, they contain charged particles called quarks that are the source of the magnetism.) In the experimental arrangement, the neutron takes one path and its magnetic field takes another. The magnetic field has a sort of out-of-body experience – at least, that's what weak measurements suggest.

The actual experiment is an adaptation for neutrons of the bomb-testing interferometer shown in Fig. 16 on p. 221. It's possible to contrive that the neutrons take 'the low road' while their magnetic fields take 'the high road', before being reunited prior to their detection. How do the experimenters know which went where on the road to the detectors? They determine this by inserting neutron absorbers in one path or the other to see if there's a dip in the output detection rate. Then look for where the magnetic fields are going using weak applied magnetic fields to see whether that alters the output. Sure enough, they see Cheshire cat states in which the neutrons go one way and their magnetism goes the other. There is an important caveat here. Remember, when weak measurements are involved, the results are statistical averages over many trials, and cannot be used to say anything about an individual particle. The separation of cat and grin can't be followed in real time. Rather, it is necessary to pick out at a later time only those neutrons that took the low road, and ignore the others in the ensemble, so the whole experiment has to be analysed retrospectively. This 'post-selection' has an aura of backward-in-time action to it, but that does not permit the experimenter to alter past events, only to deduce statistical differences in past events after the fact.

In a further variant of this peculiar idea, 'property swapping' weak measurement experiments have been performed where particles exchange attributes. It's as if you had a red bus in London and a green bus in Guatemala, which then transformed into a red bus in Guatemala and a green bus in London. Or, more theatrically, it's like you and I swapping bodies, with my identity inhabiting your body and vice versa. So not only are attributes of quantum particles not manifested until (strongly) measured, they do not even belong to the associated particles, prior to measurement! They can be passed around like tokens between different contenders.

Weak measurements have also formed the basis of many proposals for practical devices. The most advanced of these are in the field of quantum metrology (see Chapter 6): weak measurements with carefully chosen post-selection can be used to greatly amplify small differences and thus to measure certain quantities, such as magnetic gradients or light path perturbations, with enhanced precision. But the primary purpose of weak measurement theory, as with the Aspect experiment, delayed-choice, bomb-testing and eraser experiments, is to test the foundations of quantum mechanics and elucidate the nature of reality in the quantum domain.

In 1927, the evolutionary biologist J. B. S. Haldane famously wrote: 'Now, my own suspicion is that the universe is not only queerer than we suppose, but queerer than we can suppose.'[13] It is hard to contemplate the experiments I have described in this chapter without agreeing with Haldane. While physicists concur that quantum mechanics works brilliantly, there is no consensus on how to interpret the theory, even after decades of debate. Is quantum mechanics as presently formulated just a placeholder for something more intuitive and easier to grasp? Or could it point to something even weirder? Do experimentally confirmed effects like erasing the past and the influence of non-events demand a fundamental revision of our understanding of cause and effect? Does the wave function – the symbol ψ in Schrödinger's equation – represent an actual physical thing or is it just an abstract stand-in for a computational procedure? And when ψ describes blended realities, does that imply we can no longer cling to the belief in a concrete 'real world' out there, and must completely abandon the notion that there are objective facts about the physical universe? These weighty and vexing questions have plagued the theory from the outset and remain unresolved today. While they do not impede the diverse applications of the theory, they are deeply troubling if the purpose of science is to explain the world in a comprehensible way. Given these very basic concerns about the true meaning of quantum mechanics, how are we then to make sense of it?

CHAPTER 11
Making Sense of It All

The fundamental problem with quantum mechanics is that it is really two theories patched together. First, there is the wave function, which contains all that can be said about the quantum system, and which changes continuously with time in accordance with Schrödinger's equation (or its various refinements). Then there is the abrupt collapse or reduction of the wave function when something is measured, a sudden change that obeys a completely different dynamic – Born's probability rule (see p. 23). The former evolution is deterministic, but the latter is probabilistic, and the source of quantum uncertainty. As I have mentioned, quantum mechanics itself is silent on what, exactly, constitutes 'a measurement' or 'an observer', even though measurement and observation are critical to using the theory in practice. Quantum mechanics makes no sense without appending a conceptual superstructure, framed in classical physics terms, about pieces of equipment and other macroscopic objects. We've seen how attempts to rescue a hidden layer of good old-fashioned reality beneath the quantum realm – with hidden variables theories like Bohm's (p. 67) – run afoul of the experimental evidence. In my view, we have to face the fact that, down among atoms and molecules, it's impossible to say 'how things are' in the everyday sense. It won't work. So how do we patch together the real world we experience in daily life with the less-than-real quantum realm? What projects out one actual world from a ghostly amalgam of many possibilities? Does quantum mechanics fail at some level, and if so where? Or, if it's a case of quantum all the way up, how do we explain the single reality of our everyday observations? It has to be stressed that these questions remain unanswered a hundred years after they loomed large in the minds of the founders of quantum mechanics. Nevertheless, as I shall describe, there have been many attempts to lay the matter to rest.

Is the quantum world the ultimate reality?

From the outset, there was a feeling among scientists that the world of everyday experience – the so-called classical world – is the 'sensible' one and it is the quantum world that is 'weird'. To get to grips with quantum mechanics, physicists framed their approach in terms of classical concepts, and then adapted them to quantum phenomena. Quantities like energy, momentum, position, rotation and so on, which have clear meaning in daily life, get imported into the quantum realm, whereupon their properties become unusual and even startling. But they are always connected back to the classical world when a measurement is made. The very notion of 'measuring' these quantities begins by appealing to classical items of apparatus like clocks and counters and amplifiers and meters. A quantum object and the classical set-up used to investigate it form part of an indivisible whole, a point articulated most forcefully by Niels Bohr when he wrote, 'the account of the experimental arrangement and of the results of the observations must be expressed in unambiguous language with suitable application of the terminology of classical physics.' (See p. 51) Thus, the quantum world is treated as a rather bewildering extension of the classical world; it's not possible to distil out the quantum bit without considering the classical world in which it is embedded.

The description I have just given more or less corresponds to the pragmatic approach adopted by most physicists and engineers. But if you stand back and think about it, it seems to be upside down. The classical world of daily life is surely an *approximation* of the quantum world, a realm in which quantum effects are too small or too scrambled for us to notice. Pieces of apparatus, after all, are made of atoms, as are people. It seems more natural to assume that the quantum world, with all its weirdness, forms the bedrock of reality and the classical world somehow emerges from it in large or complex objects. That bottom-up approach would indeed be convincing if only one could figure out *how* the quasi-classical world of experience emerges from the quantum world. But here we hit another deep conceptual obstacle. Measurement is not just any old physical process. By definition, a measurement is an inquiry or interrogation of nature for a purpose, preconceived by a sentient being – a way to obtain specific information about the world. It is not enough to simply note that a measuring device is a (complicated) physical system in its own right; it is a measuring device because it has been carefully designed with a *function*. Geiger counters didn't just appear in the world. The Geiger counter was invented by Hans Geiger with the express purpose of detecting ionizing radiation. Be that as it may, the consideration of function or purpose gets us into very murky territory. In biology, asking what this or that feature of an organism is 'for' is fraught with philosophical problems because it seems to lead back through a chain of explanation to agency. And scientists have gone to great lengths to banish purpose from the physical world because of its mystical overtones. But perhaps in the final analysis, some form of purpose cannot be avoided when discussing quantum measurements – a conclusion that makes many scientists feel queasy.

We have seen how quantum mechanics is spectacularly successful in the microworld of atoms and molecules, but runs into issues when we attempt to apply it to systems where gravity is the dominant force, such as black holes or the cosmos. It is at this stage that a choice needs to be made. Is nature quantum mechanical at every level, right up to and including the whole universe, or does it break down somewhere? And if something changes between atom and cat, say, what is it and how does it create the classical world of daily human (and feline) experience? Could there be new physics waiting for us at the point of failure, and if so, what implications might that have for yet more transformative technology? Although most of my colleagues refuse to believe that quantum mechanics breaks down, my own feeling is that there is indeed something missing in the theory – something important – that would provide the bridge between the quantum and classical worlds.

Let's take a (critical) look at some of the proposals for what that missing piece might be.

Ways in which quantum mechanics might break down

SPONTANEOUS COLLAPSE

One suggestion is that the fuzzy quantum world effectively concretizes itself, even in the absence of measurement, thereby producing a classical state spontaneously. To simplify, it works like this: a smeared-out wave suddenly gets zapped, and – poof! – it implodes to a tiny blob. A disseminated probability wave has been transformed into a localized 'really there' particle. Maybe that could work. But what does the zapping? According to this theory, nothing; it's just random. How often does it happen? Advocates make up a number consistent with known observations. How small is the blob? Advocates make up another number (e.g. 100 nanometres). The way I've described it makes spontaneous collapse seem ad hoc and contrived, which of course it is: it's contrived to solve the thorny problem of the quantum-to-classical transition. It purports to do this by choosing the abovementioned two numbers so that isolated atoms are negligibly affected – because we do not notice anything odd at that level – but the zapping effects build up the greater the concentration of particles, so that by the time it comes to cats, wave function collapse to either live or dead is virtually instantaneous. (There are multiple versions of this theory with different features and characteristics, and I'm being rather sloppy in my description.) Spontaneous collapse theories are potentially testable in the middle ground of 'mesoscopic' objects (e.g. large molecules) which might be spotted getting zapped. To get the theory to work, Schrödinger's equation has to be mutilated to accommodate the system's vulnerability to occasional zaps.

Inelegant though it may be, spontaneous collapse has its supporters and the theory has the distinct advantage of being clear about how quantum superpositions turn into classical objects.[1] The fact that there is an underlying mechanism (zapping) does not make the theory deterministic, because the zaps are random – God still plays dice. But the price is high: two new constants of physics (the zapping rate and the blob size) have to be grafted onto existing physics. Which makes three constants in all when we add Planck's constant, which seems excessive for one theory.

What about observational consequences? Shouldn't all those 'poof!' events show up physically in some way? So it would seem. For a start, they will generate heat right across the universe, because quantum particles are everywhere and their jerky 'blobification' experiences will pervade space. Nobody has seen unexplained cosmic heating, so that puts a limit on the zapping rate. There's another idea. If charged particles get zapped, the abrupt kicks should emit photons. A European team followed up on this by going to a very dark place – the Gran Sasso Laboratory deep under the Alps – which screens out disturbances from cosmic rays. They looked for X-ray pulses that would arise from atomic nuclei undergoing jittery collapse events in a germanium crystal sheathed in copper and lead. Nothing was detected.[2] So far, spontaneous collapse theories fall into the category of 'can't rule it out but inelegant and hard to test'.

CONSCIOUSNESS TO THE RESCUE

In 1932, one of the architects of quantum mechanics, John von Neumann, analysed what happens when a quantum superposition, for example, spin-up and spin-down of an electron, is measured by a piece of apparatus such as a meter, designed so that if the electron is oriented spin-up the pointer on the meter goes left and if it is spin-down the pointer goes right. If the apparatus is also regarded as a quantum system, then when it is coupled to the electron, the up-down superposition state of the electron becomes entangled with the state of the meter. The composite system, electron-plus-meter, is now in a superposition of states: there has been no collapse of the wave function, and the pointer will be in an ambiguous blended state of left-and-right. If the electron-plus-meter is now coupled to a larger piece of apparatus designed to measure which way, left or right, the pointer is oriented, then once again the total composite system is thrown into a superposition of states. There thus seems to be an open-ended chain; at no point does a final result get projected out of the amalgam. Where is the elusive Heisenberg cut (see p. 52)? Von Neumann suggested that to avoid an infinite regression, consciousness might play a role. Maybe the wave function finally collapses when it enters the mind of an observer?

In an extension of this line of reasoning, Eugene Wigner envisaged a friend in a laboratory (substituting for a cat in a box) performing the quantum spin-up/down experiment. Following von Neumann's proposal, when the friend (presumed conscious) measures the spin, the electron's superposition purportedly collapses into one of the definite states (spin-up or spin-down). From the friend's perspective, the measurement has been made, and the electron is in a definite state. However, Wigner, who is outside the lab and unaware of his friend's measurement outcome, must still describe the entire lab (including the electron and his friend) as being in a superposition. From Wigner's viewpoint, the composite system of friend plus electron is entangled in a superposition of all possible outcomes until Wigner himself measures or interacts with the system in some way, such as by walking into the lab and interrogating his friend about their experiences. All of which seems bizarre.

The mind-over-matter resolution of the measurement problem is widely rejected (and indeed was later abandoned by Wigner himself). Scientists are generally ill-disposed to incorporating consciousness into physics in any capacity. Like spontaneous collapse, consciousness collapse looks ad hoc and contrived. What, exactly, constitutes a superposition-busting mental state? Does it require a human brain (Wigner's specifically), or will a cat do its own 'I'm alive' observation in Schrödinger's notorious thought experiment? How about a cockroach? A bacterium? These deliberations raise deep questions about the nature of consciousness and observation in quantum theory.

There is another issue here. Are we to suppose that the entire cosmos was in a state of suspended superposition for billions of years until along came sentient beings who started annihilating wave interferences, turning boths into either/ors? That was indeed more or less the position argued by John Wheeler, based on his delayed-choice experiment that I discussed in the previous chapter (see pp. 227–30). Wheeler's point was that even if the universe had to wait aeons for the emergence of minds (e.g. human beings) to sharpen quantum blurriness, their observations could nevertheless reach back in time and concretize, i.e. snap into definiteness, at least some fine details of the pre-sentient cosmos.

He termed this an 'existence loop' in a 'meaning circuit' and referred to the set of ideas as 'the participatory universe', in which conscious beings become co-creators (or at least co-concretizers) with the very universe that gave rise to them.[3]

Most physicists regard the participatory universe to be loopy (in both senses), and shy away from any attempt to invoke consciousness in quantum measurement. But in defence of von Neumann, Wigner and Wheeler, there is a general point to be made. If the job of science is to explain the world, then it cannot evade consciousness, because conscious experiences are undeniably part of observed reality. The key question is whether consciousness is a fundamental physical phenomenon or merely an embellishment, an incidental quirk of biology (an 'epiphenomenon', to use the philosopher's jargon). And if the former, is it a quantum phenomenon? If you believe the world is thoroughly quantum at all levels, then quantum mechanics has to be able to accommodate conscious experiences. There has to be some aspect of a quantum state you can point to as the quantum correlate of this or that mental state. Nobody has done this.

But that is not quite the end of the matter, for there is an obverse to the proposal that consciousness causes the wave function to collapse: perhaps the collapse causes consciousness.

GRAVITY IS THE CULPRIT

At the end of Chapter 9 I explained how physicists have struggled to incorporate gravitation into quantum mechanics. Unlike the other forces of nature (electromagnetism and the two nuclear forces) which have credible quantum descriptions, gravitation is the odd one out. If, then, gravitation has a unique status in the quantum realm, perhaps it is the facilitator of wave function collapse? After all, gravity affects everything, so it is a candidate for an all-pervasive process. Furthermore, the gravity of an object grows with its mass, so we can imagine that a macroscopic object will be classical whereas an atom will be fuzzy and quantum. But just how massive must an object be to undergo wave function collapse? An answer has been provided by Roger Penrose and his collaborator Stuart Hameroff. Their theory differs from other spontaneous collapse theories in a crucial respect.

Here is an outline of the Hameroff–Penrose proposal. Suppose you have a quantum object in a superposition of a particle here and a particle there, with a certain separation. If these were real, classical particles, there would be a (tiny) gravitational attraction between them, with an associated energy – the energy you could extract by allowing the particles to fall together under the force of their mutual gravitational attraction. Pretend now that the quantum superposition of ghost-particle-here and ghost-particle-there is treated the same as if they were two identical real particles at those locations, that is, there is an energy cost associated with the two objects being pulled apart. Then according to Penrose and Hameroff, gravity will try to collapse the superposition onto one or other of the two positions for the object on a time scale determined by Planck's constant, in a formula reminiscent of Heisenberg's energy-time uncertainty principle (see p. 169): the bigger the gravitational energy, the shorter the time. The numbers look sensible. Electrons have such a small gravitational effect that they might stave off collapse for thousands of years, whereas the legendary cat would be reduced almost instantaneously into either an alive or dead state. Penrose and Hameroff developed this theory in part as an explanation for consciousness, by considering what happens when the collapse events occur in the brain. In their words: '. . . the "choice" involved in any quantum state-reduction process would be accompanied by a (minuscule) proto-element of experience, which we refer to as a moment of *proto-consciousness*'.[4] This theory thus does the double duty of recovering the classical world from the quantum by providing an explicit physical mechanism for collapse, and also by incorporating what it means to be conscious into a unifying physical theory.

I THINK THEREFORE I COLLAPSE THE WAVE FUNCTION

As I've explained, some people think consciousness might cause wave function collapse and others think wave function collapse might cause consciousness. But there is a third possibility. Could it be that X causes consciousness and X also causes the wave function to collapse to a classical concrete state, but consciousness per se doesn't *cause* the collapse (or vice versa)? Instead, both consciousness and collapse stem from a common cause: X. What is X? Theories of consciousness are many and murky, but most of them accept that the brain has something to do with it, and the brain is the most complex physical system we know. Accepting that consciousness is an emergent property of complexity, a physical system would have to be sufficiently complex to generate sentience or agency (in some sense, still vague). If the foregoing line of reasoning is correct, then quantum mechanics breaks down not due to physical size (superconductors can be large and are still quantum) or mass (interference has been observed with large molecules and even mesoscopic cantilevers and membranes, not to mention the multi-kilogram mirrors at $LIGO$) but *complexity*. A small molecule like ammonia is too simple for quantum mechanics to go awry, but a cat is complex enough to be on the classical side of the transition. Somewhere between ammonia and cat, quantum mechanics breaks down, and consciousness kicks in. Could it be that quantum mechanics is merely an *effective* theory that works well (it is certainly effective!) in low-complexity situations, but needs to be replaced by a better theory for general systems of arbitrary complexity? Physicists are familiar with the concept of effective theories that work well enough at low energies, but must be replaced at high energies by something more comprehensive. What I'm proposing is a similar idea, with complexity rather than energy being the relevant parameter – though I am far from the first person to suggest this. For example, the Nobel Prize winner Anthony Leggett wrote thirty years ago, 'It is quite conceivable that at the level of complex, macroscopic objects, the quantum mechanics superposition simply fails to give a correct account of the dynamics of the system.'[5]

The obstacle to refining this explanation is to come up with a plausible quantification of complexity. Complexity has many competing mathematical definitions, but as it happens there is one that is deliberately chosen to capture some features of consciousness. Called integrated information, it is a measure, not of the *amount* of information in a system, but the way it is organized, the way it flows around the system in feedback loops. I don't want to get into the technical details here, and in any case, it may turn out that integrated information is not the best characterization of what is going on in a brain. But it is a good first try. The point is that some combination of neural architecture and information traffic must underlie consciousness – perhaps 'agency' is a better concept – and that same measure just might also underlie the resolution of the quantum measurement problem. The natural way to test this hypothesis is to build quantum information networks of increasing complexity (by some definition) and look for the emergence of classical behaviour. (Technically speaking, one would look for departures from unitarity.) That would not just involve a head-count of qubits, but also take account of how they are strung together. I'm keeping my fingers crossed that quantum computers, when they get a bit bigger and more complex, will uncover something new and surprising along these lines.

EVEN SPOOKIER?

A similar set of ideas – that quantum mechanics is merely an effective theory so far investigated under a limited set of circumstances – goes under the cryptic name of 'PR boxes', after its proponents Sandu Popescu and Danny Rohrlich. What Popescu and Rohrlich suggest is that quantum mechanics is a placeholder – a half-way house to a future super-duper theory of the world that recovers quantum mechanics as an approximation, and which might also clarify what a measurement/observation could be, that is, what causes the wave function to collapse. It's intriguing to speculate that if there is a bigger theory (in some sense) in which quantum mechanics is embedded, it is likely to be even weirder than quantum mechanics. Weirder how? Perhaps spookier.

Earlier, I explained how two quantum particles far apart can nevertheless stay in a sort-of telepathic contact, although they can't exchange actual information. Einstein famously called this 'spooky'. An interesting question is whether the world could be spookier – that is, could the telepathic link be even stronger than that predicted by quantum mechanics? Popescu and Rohrlich investigated this question in the context of another simple cooperative Alice-and-Bob game like the magic square described on p. 78. It goes roughly like this.[6] (Readers suffering from Alice-and-Bob burnout might want to just take my word for it and skip ahead.) Imagine a pair of 'black boxes', each with a slot in the top and a light on the side that will flash either red or blue. There is a pile of red tokens and blue tokens, equal numbers. Tokens can be dropped one by one in the slot, and each token makes the light flash once. For example, in goes a red token and the light flashes blue. The referee, Charlie, draws tokens from the pile randomly and blindly assigns one to each player. He informs Alice and Bob that the rules for jointly winning a prize are that if they both drop in a red token, their box's lights flash different colours, but in the three other combinations of input (blue-blue, blue-red, red-blue) the lights must flash the same colour (either red or blue) to win. Alice and Bob, clever physicists that they are, are challenged to design the boxes' innards so as to maximize their winnings, on the understanding that when the game starts they will remain in different rooms and won't communicate.

How well can they do? If the lights just flashed randomly, Alice and Bob would win on average only half the time. But they can do better, and achieve three out of four on average, by cleverly rigging their boxes with colour readers and wires. For example, they can ensure that whenever they each put in a blue token they both get a blue light. But suppose Alice puts in blue and Bob red? Bob needs his box to flash blue that time too, to win. He could cover both the foregoing eventualities and rig things to *always* flash a blue light. If Alice goes for a red token and Bob blue, a win demands that Alice's light is blue; well, she too can rig her box for that to happen. Those box designs give three wins out of the four input-colour combinations. However, the pair will lose a round if they are both given red tokens by Charlie, because they will then both get a blue light, which breaks the 'different colour' rule for red-red. No matter how you rig up the boxes' workings, you can never exceed that ¾ success rate.

Nevertheless, because Alice and Bob have read the earlier sections of my book (especially the one dealing with the magic square), they will know they can beat the ¾ limit by using quantum entanglement. If, instead of clunky mechanical contraptions inside the boxes, they each store one of an entangled pair of photons, carefully protected against decoherence, then they can exploit spooky pseudo-telepathy to win more often. Unlike for the magic square example, however, quantum entanglement can't deliver 100 per cent success for the two-box game; it turns out the best one can do (following some mathematics that I shall omit here) is about 85 per cent success. Which is a very deep result. What it says is, sure, quantum mechanics implies weird telepathic correlations between widely separated particles, but quantum mechanics isn't *maximally* telepathic. One could imagine even spookier correlations that would ensure success for Alice and Bob *every time* they play a round of the two-box game. That would not be quantum mechanics; it would be some yet-to-be-discovered post-quantum mechanics. Might the world in fact be like that, and we haven't noticed?

If we don't actually have a theory of maximally spooky post-quantum mechanics we are a bit hamstrung in investigating the matter. But even without a full theory, some things can be checked out just on the basis of the assumed excessive at-a-distance correlations. For example, could a spookier computer offer even more thrilling prospects than a quantum computer? How might the additional spookiness transform the prospects for long-range communication of information? Could a spookier internet outperform the quantum internet, or a super-spooky pair of investors outperform the stock index? And is there a snag? Do the extra correlations open the way to faster-than-light communication, in violation of the theory of relativity? These matters are under investigation.

In addition to the tantalizing technological possibilities, there is a philosophical issue here. If it transpires that some form of spookier post-quantum mechanics is possible in nature without violating the theory of relativity, and if it turns out that we do *not* live in such a world, then why has nature opted for some spookiness, but not gone the whole hog? After all, if spookiness is good, surely more is better? Is there something special about okay-as-far-as-it-goes spooky quantum mechanics, which singles out *that* theory from a whole spectrum of possible theories (on a sliding scale of spookiness) up to the maximum? If so, what? Nobody knows.

No breakdown: quantum mechanics should be accepted as it is

The foregoing survey of quantum breakdown proposals is far from complete, but I now want to turn to a different set of interpretations, which assume that quantum mechanics appears bizarre and paradoxical not because it is wrong or incomplete, but because we are thinking about the world the wrong way.

WHY WE DON'T SEE LIVE-DEAD CATS

We can put an atom into a superposition of two distinct states, for example, atom here, atom there. But on a macroscale, we don't see such superpositions. I just checked: the clock on my desk is sitting at a single definite location – the same place I moved it to yesterday. If quantum mechanics applies all the way up, why aren't there 'Schrödinger-cat' states all around us? There are in fact two puzzles in one here. The first is why micro 'both' goes over into macro 'either/or'; e.g. from live/dead cat to live cat *or* dead cat. The second is which outcome actually happens – live or dead? After some decades of bafflement, the first puzzle is now solved. Cats – and almost all macroscopic physical systems – are continuously coupled to their environment by countless interactions, such as from photons and air molecules impinging on them. This barrage of disturbances quickly establishes quantum entanglement between parts of the object (cat, clock . . .) and a vast number of randomly distributed atoms and molecules surrounding it. As a result, any initial coherent superposition in the state of the macroscopic object is *very* rapidly scrambled and effectively eliminated from any conceivable observation.[7] So, as a *practical* matter, one can safely ignore the infinitesimal remaining overlap between feline aliveness and deadness: an observation would always yield either alive or dead. Problem solved! But not so fast: the issue of amalgamated realities hasn't totally gone away by invoking decoherence: it is still lurking there in the total system, which includes the wider environment, albeit in a hopelessly scrambled form. (Recall the problem of von Neumann's 'chain': see p. 253.) And the second problem – why alive rather than dead, or vice versa – also needs addressing. Why, among the various alternative realities, is one particular result projected out? How does nature pick a particular winner in the great cosmic lottery? Decoherence alone doesn't answer that.

SUPERDETERMINISM

Einstein didn't like a dice-playing deity. Nor do some contemporary physicists such as Sabine Hossenfelder, the well-known science commentator, who thinks that randomness at the heart of nature is a horrible idea.[8] But haven't we seen how Bell-type experiments have discredited the notion of an underlying deterministic theory involving a sea of mysterious, yet-to-be-discovered, hidden variables? Well, not totally. Recall that the power of the experiments on entangled particles stems from the possibility that the experimenters can change their minds suddenly at the last moment. What if it were the case, however, that what the experimenters actually choose is also determined in advance by hidden variables, then the entire system of particles, apparatus, laboratory and observers – indeed the entire universe – might be nailed down in *every* detail from time immemorial in a vast cosmic conspiracy? Alice and Bob are simply fated to do what they do and measure what they measure. More precisely, what happens at A is not statistically independent of what happens at B because they are all part of the Great Plan. This theory is often derided on the basis that it assumes a vast cosmic conspiracy set in train at the origin of the universe. However, while Alice and Bob may feel that they are free to choose their polarizer angles in a Bell experiment, they can do so to only a finite accuracy, leaving a loophole for hidden variables to determine the actual precise angles. Oxford University physicist Tim Palmer has proposed an explicit model of such 'superdeterminism without conspiracy' along these lines, in which he argues that a hypothetical world where the pair chose otherwise is forbidden by the (unknown) laws of hidden variable physics.[9] In summary, superdeterminism is an attempt to keep quantum mechanics intact but reconcile it with a form of objective reality – albeit in a rather subtle way.

ALL IN THE MIND?

Quantum mechanics is rooted in concepts like uncertainty, indeterminism and probability. Most people have a rough understanding of probability as how likely something might be to happen, but what does that really mean? Mathematicians and philosophers have spent centuries squabbling about it. Is the probability of an event or outcome a property of the world or a property of measurement or a characteristic of what people believe to be the case? Probability is just as riven with interpretational issues as quantum mechanics. So maybe the bafflement about the quantum measurement problem (see Box 3) simply reflects the muddle over what to make of the probability concept. If you flip a coin many times then you get heads for about half the sample; the more trials you conduct, the closer to 0.5 you get. In that case, probability has a fairly clear meaning in terms of the frequency of the particular outcome. But what about a single trial? You flip a coin once. In quantum mechanics one often does a one-shot measurement and the wave function is supposed to tell you the probability for that one individual case. What does that even mean?

Perhaps a better way to think about probability goes back to an eighteenth-century English clergyman named Thomas Bayes. Roughly speaking, this view treats probability as the degree of belief. If something is very probable, you are justified in believing it is the case. If you hold a prior degree of belief about something and you update your information about it by an observation, then your degree of belief may change. We often use Bayesian reasoning in daily life. For example, you think it might rain today based on the weather forecast, but these are notoriously unreliable, so you judge there is a 50 per cent chance of rain. Thus, your Bayesian prior is 0.5. Then you look out of the window and see huge thunderclouds, so you update your belief in rain to, say, 0.9. Maybe quantum states are just degrees of personal belief – the wave function being the Bayesian prior, and the collapse of the wave function on measurement is the update. Ultimately it is all in the mind of the beholder of what is a reasonable and rational expectation to hold about the world. That point of view is called quantum Bayesianism, or, more succinctly, Qbism, and it has some enthusiastic proponents.[10] Qbism, in effect, shifts the measurement problem from physics to the realm of personal experience and judgement. Is that progress? Or just moving the bump in the interpretational carpet? To a hard-nosed physicist, Qbism has the air of a philosophical whitewash that buries a perplexing physical process in a lot of sophistry. It doesn't solve the measurement problem so much as define it away.

FACTS ARE RELATIVE

An alternative approach starts with the question: is the physical universe a collection of objects that interact, or just a set of relations, from which we find it convenient to attribute properties (mistakenly) to actually existing things? Relational quantum mechanics (RQM) is based on the idea that *all* physical properties are relational, meaning that objects are defined solely in terms of the interactions between physical systems rather than as inherent properties of individual systems themselves. And it's true that you can't totally isolate anything from its environment. An electron cannot be separated in real time from its electromagnetic field, for example (Cheshire cat states notwithstanding), and its properties like electric charge can be observed only through its interaction with an external system. According to RQM, observers are physical systems that are entangled with other systems, and the information they acquire through measurements is entirely relational to that observer. A strong proponent of RQM, Carlo Rovelli, puts it like this: 'The world that we know, that relates to us, that interests us, what we call "reality", is the vast web of interacting entities, of which we are a part, that manifest themselves by interacting with each other.'[11] Again, there is a shift from what *is* in the world out there to how an agent or being relates to the world and what that agent can and cannot know through the process of measurement. In this picture, there are no absolute facts about the world, only facts relative to whatever (or whoever) is interrogating nature. A closely related set of ideas has been explored by Robert Spekkens and David Schmidt at the Perimeter Institute in Waterloo, Canada, and John Selby of the University of Gdansk in Poland, who endeavour to use new forms of mathematics to unscramble 'Jaynes's omelette' (see p. 54) – the confusing quantum mixture of *physical* causation (what causes what in the outside world) from an observer's *mental* chain of inference (how our observations alter what we think to be the case).[12]

One shortcoming of RQM, in my opinion, is that our concepts of space and time are abstractions we derive from comparing the results of measuring devices made of matter: how far apart are two objects? What is the duration between two events? If it is merely relationships that lie at the base of our world view, then space and time (more accurately spacetime) has no independent existence. But the entire field of relativity and cosmology is founded on the notion that spacetime is a physical *thing* that, even in the absence of matter, has its own dynamics. I concede that my position could be overly simplistic; spacetime might itself be built out of some quantum-y substructure, but I'd like to see that worked out first before changing my mind.

UNKNOWN OR UNKNOWABLE?

In the early 1930s, when physicists were struggling to understand the nature of quantum uncertainty and what can in principle be known about a quantum system, an equally profound revolution burst upon the world of mathematics and the foundations of logic. In 1931, the Austrian logician Kurt Gödel proved that mathematics contains statements that are in principle undecidable – that is, it cannot be determined whether they are true or false. Gödel based his argument on the axiomatic method of reasoning, in which certain statements (axioms) are accepted as self-evidently true (for example, when zero is added to a number the resulting number is unchanged) and then the rules of standard logic are used to deduce statements about numbers. But if a combination of axioms and logical deductions cannot answer certain propositions about arithmetic, then in a sense, mathematics is always incomplete, however much it is augmented by new rules and definitions. Later, Alan Turing showed that there exist numbers that are uncomputable, even allowing unlimited time for a computer to chug away and eventually halt at the answer (see p. 101). There is no general procedure to tell in advance whether the machine will in fact halt. The work of Gödel, Turing and Alonzo Church proved that there is a fundamental source of unknowability in the rational foundations of existence. By extension, the laws of physics, which are mathematical relationships, inherit this inescapable restriction. Wheeler was fond of asking the question, 'How come the quantum?' Towards the end of his life he became convinced that there is a deep link between the mathematical unknowability of Gödel and the physical unknowability of the quantum universe.[13] Perhaps when the physicist asks a question like, 'Is the spin pointing up or down for this quantum state?' and has to confront that, in general, the answer is unknowable in advance, it is analogous to asking whether a certain mathematical proposition is true or false. Attempts to derive quantum mechanics from foundational axioms have to factor in that Gödel's theorem challenges the whole notion of axiomatic reasoning. While I am attracted to Wheeler's hunch, it is very far from being worked out in detail. It remains what Wheeler himself would have described as 'an idea for an idea'.

AN EMBARRASSMENT OF UNIVERSES

I have saved the currently most favoured interpretation of quantum mechanics for last. Suppose we have a quantum state in which an atom is in a superposition of over-here and over-there. In the absence of measurement, we cannot say for sure where it is. But, as we've established, following a position measurement, one of these contending ghost realities abruptly vanishes, and the quantum state jumps immediately into either atom-here or atom-there: the wave function collapses.

What I have just described is more or less the conventional understanding of the effect of an observation in quantum mechanics, in which the act of measurement promotes a ghostly 'reality contender' into the real deal, and simultaneously annihilates the competition. But there is a radical point of view, gaining in popularity, which views the outcome of the measurement not as either/or, but *both*. That is to say, following the measurement, there is no 'collapse', no 'picking the winner'. Rather, each potential world becomes a real world, one with an atom here, the other with an atom there, existing in parallel with each other. But, you might object, we see only one reality. How can there be two? The answer lies in the 'we'. In the above so-called many-worlds interpretation of quantum mechanics, each of the two parallel worlds contains a version of you and me, identical apart from the experience of atom here versus atom there. (For this reason, the theory is sometimes called the many-minds interpretation.) By extension of the same basic idea, Schrödinger's cat is observed to be both alive and dead: alive in universe *A*, dead in universe *B*.

The many-worlds theory was formulated in 1957 by Hugh

Everett III, a physics graduate student. For a long time, his idea had few takers, but over the years it has grown in popularity, until I would say it is now the majority view among theoretical physicists working on foundational issues. (I attend conferences at which I would be a bit embarrassed to admit that I am not a fan of Everett's interpretation.) One reason for this widespread acceptance is the theory's austere simplicity. Quantum mechanics is taken at face value. The wave function, with its myriad components, each representing a possible world, *is* the reality. Nothing is added about observers choosing what to measure or collapsing possibilities or contending loser worlds that vanish without fanfare. No funny extra mechanism kicks in to turn a superposition into a single state of affairs. All possibilities are there at once, forming an infinite collection of parallel universes, often termed a multiverse. Part of the appeal of the many-universe interpretation comes from the subject of quantum cosmology, in which one discusses a wave function for the whole universe. Since there is nothing outside the universe – by definition – to effect a collapse, then all possibilities must co-exist in parallel. If you believe quantum mechanics is the theory of the world, then there is no room in the many-worlds theory to add anything to project into existence the reality of daily life. You just have to accept an infinity of realities, side-by-side.

Some people are appalled by the thought that there will be an infinite number of duplicate beings. 'Which one is the real me?' they ask. 'Each of them is a real me, from a "that me" point of view,' is the answer. And for every me, there will be a vast number of nearly-mes, and an even bigger number of sort-of-mes, and so on. Personally (pun intended), I don't have a problem with many mes. But I do have reservations about the Everett interpretation, because I take consciousness seriously. The act of observation provides a nexus between conscious agency and the external universe, which is swept away in the many-universes interpretation; nothing distinguishes a branching of realities due to a mindful measurement from a mere splitting due to any random interaction. So we have lost the opportunity to explain what is for many people the most important and primitive fact about existence – namely, their own conscious experience.

Box 6

Parallel universes

If everything is quantum, up to and including the entire universe, then the whole ghostly paraphernalia of the quantum world must persist on the macro-scale. For example, just as an atom can be in many places at once, it would then be the case that the Earth – or, for that matter, you or I – will also be in many places at once. How can that be? We only ever observe one Earth in one location. A possible answer is to invoke parallel realities, or multiple universes. For example, suppose one atom collides with another and bounces off. A standard undergraduate exercise is to use quantum mechanics to calculate the relative probabilities that the ricocheting atom will be found heading in any particular direction.

In the Copenhagen interpretation, there would be a definite unique fact of the matter that would spring into concrete existence when the scattered atom is detected in some way. But Hugh Everett proposed that *all possible* trajectories are equally real, each one existing in its own complete universe. Furthermore, many such universes would come with observers who might well assume that their universe was the only reality. In fact, there would be many realities inhabited by many versions of observers – many universes and many minds, if you like. Inevitably, Everett's argument leads to an infinity of parallel realities. It's expensive on universes, but cheap on epistemology.

Can anyone glimpse the other worlds?

Fans of many universes point out that their interpretation is superior because nothing is added to quantum mechanics. You don't need to jump through mental hoops to decide which universe is 'The One', projected out of the infinite stack of possibilities on offer, or how it came to be promoted to reality by some sort of 'observation' or a random spontaneous collapse.

Critics counter that the many-universes theory is untestable – what evidence is there that all those parallel universes exist? We only ever see one. Forty years ago, I made a BBC radio documentary about quantum reality called *The Ghost in the Atom*.[14] This was long before Everett's theory took off, but an early devotee was David Deutsch, he of the quantum computer. I asked him how the theory could be tested. Here is a transcript of our (somewhat light-hearted) conversation, edited for brevity:

> **Deutsch**: Here's how it works: we first find a situation in which the conventional interpretation predicts that all the other universes suddenly disappear, and where the Everett interpretation predicts that they don't disappear but they're all there in parallel. Then we find some observable consequence of their subsequently interacting with each other in an interference experiment. And we then observe one result if the Everett interpretation is true, and another result if any of the conventional interpretations is true. Simple as that.
>
> Unfortunately, this experiment requires the observation of interference effects between two different states of an observer's memory. So the place where we would expect a crucial experimental test is with quantum effects inside an observer's brain.

(The word 'brain' here is being used generically to refer to any physical system capable of generating conscious experiences and storing memories, not specifically an organ like the human brain.)

> **Davies**: We're talking about quantum memory?
>
> **Deutsch**: We're talking about quantum memory, and presumably electronic artificial intelligence.

Deutsch is talking about quantum AI – 'quintelligence' – all those years ago! To continue:

Davies: But we can envisage building some sort of artificial superbrain with a memory at the quantum level, and ask it to carry out this experiment for us, and tell us what it feels?

Deutsch: That's right. And it could record the results of this experiment in any way we like. It could perhaps write them down, or tell us the results; the difference – rather like in Aspect's experiments – between quantum theory and the rivals is not a matter of a small percentage, it is an all-or-nothing thing. In the experiment I describe, one would observe a certain atomic spin, and if it was pointing one way, Everett's interpretation would be true, and if it was pointing the other way, the conventional [Copenhagen] interpretation would be true.

Davies: Exactly what experiment does he perform, if we can call him he?

Deutsch: Yes. The experiment hinges on observing an interference phenomenon inside the mind of this artificial observer. This can either be done by someone else looking inside him, or more elegantly, by his trying to remember various things so that he can conduct an experiment on his own brain while it's working.

Davies: He can observe himself?

Deutsch: He can observe part of himself, yes. And what he tries to observe is an interference phenomenon between different states of his own brain. In other words, he tries to observe the effect of different internal states of his brain in different universes interacting with each other.

Davies: How would these different internal states be set up?

Deutsch: They are set up in the first instance by a special sense organ which is essentially just another quantum memory unit. This sense organ is used to observe the state of an atomic system – a system with two possible states, such as an atomic spin, for example. Now, quantum theory predicts that, having observed this atomic system, the observer's mind will differentiate itself into two universe branches [i.e. branches of the wave function].

Davies: So, we have an atomic system with two possible states, each of which would trigger the brain of this artificial observer to be in either one state or the other. But we don't let the universes get out of touch with each other. We bring them back to overlap, to interfere with each other, and this poor observer is, as it were, schizophrenic and observing both of these possibilities at once.

Deutsch: That's right. In effect he is feeling himself split into two copies.

Davies: And he feels himself merge again?

Deutsch: Yes, in effect. Of course, we don't have sense organs of this type, so it's hard to say what this would feel like, but when this observer exists we can ask him!

Davies: It sounds most uncomfortable!

Deutsch: Perhaps it will be, but then presumably he'll be a physicist so he'll enjoy doing this experiment!

Obviously, we don't experience 'the universe next door' bleeding into ours in this manner. That's because humans are macroscopic systems in which decoherence is rapid and serves to decouple the different components of any quantum superpositions faster than the speed of thought. (Though, recall the claims of quantum biology, and Penrose's work in particular, that contest this.) But just because the other universes have swiftly got out of touch with 'this one' doesn't mean they don't exist, it just means they have slipped beyond the grasp of we mere mortals, but perhaps not beyond the grasp of any forthcoming quantum AIs.

The possibility of *quantum* observers as opposed to (classical) observers (like us) of quantum systems goes right to the heart of the nature of reality – what philosophers call ontology. The Copenhagen interpretation dodges explaining, or even asking, 'what is really going on' in a quantum system, and talks instead about what observations and measurements reveal. Niels Bohr expressed it succinctly: 'It is wrong to think that the task of physics is to find out how nature is. Physics concerns what we can say about nature.'[15] An observer like us sees one world, one version of reality, an infinitesimal subset of what is in the wave function. If classical observers are all there are or ever can be, then reality is indeed a projection of one world from an infinity of potential or possible worlds. But if there can exist quantum observers of the sort Deutsch long ago envisaged (technically, minds that inhabit Hilbert space rather than mere three-dimensional physical space) then the potential mental realm opened up is exponentially larger than we humans can ever experience, because it would encompass all possible worlds. Of course, some visionary writers predict a future in which human minds are augmented by quantum consciousness (a qubit processor implanted in the brain, say) or uploaded into quantum computers. (Recall my discussion of brain-computer quantum interfaces and use your imagination.) Then these superminds would have the entire Hilbert space to explore.

It's fascinating to speculate what would happen if we built a quantum AI and asked it to solve the quantum measurement problem. (For that matter, even a classical AI might surprise us with something novel.) If Deutsch is right, the quantum AI would reply: 'There is no problem: everything is quantum'. In which case we would ask it: 'Then what, precisely, is a measurement?' Or, more daringly, 'What is an observer?' The reply to the latter might be, 'I am. But you, I'm sorry to say, are a diminished and incomplete specimen of an observer.' What I suspect, however, is that the quantum AI would articulate a theory of observation that would be completely incomprehensible to us, and mark the end of humanity's glorious quest to understand existence based on our biological intellectual resources alone. I hope that won't be the case, but there is no law to say, even though we have been incredibly successful in unravelling the secrets of nature, that the project is completable. We might simply get stuck and concede defeat, handing over to the awesome intellectual giant that quantum AI would represent.

Our quantum future

I could go on, but won't. I have only scratched the surface of the many competing nuanced positions on the interpretation of quantum mechanics. It's been nearly a century and scientists are still bickering over the meaning of the theory, which indicates how deep and troubling the implications are for the nature of reality. I'm not really a bickering sort of person, so I don't take a strident position on the matter. I used to be a fan of the many-universes interpretation when it was still novel and unpopular, but then I drifted away from it when most of my colleagues piled in, partly because of my contrarian nature, but also because, as I have explained, I take mind seriously and most versions of many universes attempt to eliminate consciousness from the story. These days, I'm inclined to believe that quantum mechanics as currently formulated is a so-called effective theory that breaks down in more complex systems in the manner I outlined on pp. 257-9.

It seems likely that advances in quantum technology will open the way to testing some of the models described in the foregoing pages. For example, if Penrose is right about gravitational effects collapsing the wave function, then it should be possible to detect this in the laboratory with sensitive enough equipment, and indeed several attempts are being planned. If, on the other hand, quantum mechanics starts to break down at a certain level of complexity, it might be possible to demonstrate this in the realm of 'mesoscopic' molecular systems, for example, in organic molecules like ATP, which has 67 atoms and is the powerhouse for many of life's processes. The challenge – as always – will be to screen out decohering environmental disturbances. Another increasingly complex system is the quantum computer, because qubit entanglement is exponential. Consider that a mere 300 qubits can be paired in about 10^{90} ways, which could be used to form a quantum superposition with more components than all the particles in the universe. As the number of qubits increases in the coming years, quantum computers will prove to be one of the most stringent tests of quantum mechanics, assuming they operate as predicted.

A major unfinished line of research is quantum biology. The difficulty of studying quantum effects in systems as complex as living organisms may be concealing a rich variety of quantum phenomena critical to understanding how life operates; which might also provide major technological advances through biomimicry. We must, however, always be open to the possibility that a new type of experiment will lead to a discovery that could transform the subject overnight, perhaps uncovering a 'post-quantum' regime of the sort I touched on earlier in this chapter.

One of the deepest questions that scientists ask is: why is the world the way it is? Humans didn't invent quantum mechanics: nature gave it to us. To fully grasp the quantum aspects of the world entails struggling with both technical details and, more challengingly, with profoundly alien concepts. And yet quantum mechanics is extraordinarily beautiful and is one of the most secure branches of human knowledge, even though its completion remains elusive. It is a joy to realize that, as far as we know, this mysterious world of the quantum ultimately underpins everything we experience, connecting the abstract beauty of mathematics to the tangible phenomena around us, and linking the lowliest subatomic particle to the grandeur of the cosmos. To be aware of the quantum world is to glimpse something of the majesty and elegance of the physical universe and our place within it.

And so, we leave this tantalizing subject curiously unfinished, a job for the next generation of physicists.

Bibliography

Decoherence and the Quantum-to-Classical Transition by Maximillian
Schlosshauer (Springer, 2007).

The Fabric of Reality by David Deutsch (Allen Lane, 1997).

The Ghost in the Atom: A Discussion of the Mysteries of Quantum Physics,
P. C. W. Davies & Julian R. Brown, eds (Cambridge University Press, Canto
edition, 1993).

Helgoland by Carlo Rovelli (Allen Lane, 2020).

The Hidden Reality: Parallel Universes and the Deep Laws of the Cosmos by Brian
Greene (Knopf, 2011).

How the Hippies Saved Physics: Science, Counterculture and the Quantum Revival
by David Kaiser (W. W. Norton, 2011).

*Incompleteness, Nonlocality, and Realism: A Prolegomenon to the Philosophy
of Quantum Mechanics* by Michael Redhead (Clarendon Press, reprint
edition, 1989).

Life on the Edge: The Coming of Age of Quantum Biology by Johnjoe McFadden &
Jim Al-Khalili (Crown, 2015).

Niels Bohr: A Centenary Volume by Aage Petersen (Harvard University
Press, 1985).

Oxford Handbook of the History of Quantum Interpretations, Olival Freire, Jr, ed.
(Oxford University Press, 2022).

Philosophy of Physics: Quantum Theory by Tim Mauldin (Princeton Foundations
of Contemporary Philosophy, 2019).

Portals to a New Reality by Vlatko Vedral (Basic Books, 2025).

*The Primacy of Doubt: From Quantum Physics to Climate Change, How the Science
of Uncertainty Can Help Us Understand Our Chaotic World* by Tim Palmer
(Basic Books, 2022).

Quantum: A Guide for the Perplexed by Jim Al-Khalili (Orion, 2012).

Quantum Computing Since Democritus by Scott Aaronson (Cambridge
University Press, 2013).

Quantum Information Theory and the Foundations of Quantum Mechanics by Christopher G. Timpson (Oxford Academic, 2013).

The Quantum Internet: The Second Quantum Revolution by Peter P. Rohde (Cambridge University Press, 2021).

Quantum Measurement: Theory and Practice by Andrew N. Jordan & Irfan A. Siddiqi (Cambridge University Press, 2024).

Quantum Mechanics and Experience by David Z. Albert (Harvard University Press, 1992).

Quantum Paradoxes: Quantum Theory for the Perplexed by Yakir Aharonov & Daniel Rohrlich (Wiley, 2005).

Quantum Physics for Dummies by Steven Holzner (revised edition, Wiley, 2013).

Quantum Supremacy: How the Quantum Computer Revolution Will Change Everything by Michio Kaku (Doubleday, 2023).

'Quantum Technologies: On the Cusp of a Revolution', *Physics World* 34, No. 12 (2021).

The Quantum Universe: Everything That Can Happen Does Happen by Brian Cox and Jeff Forshaw (Penguin, 2012).

The Quantum World: Quantum Physics for Everyone by Kenneth W. Ford (Harvard University Press, 2004).

Schrödinger: Life and Thought by Walter J. Moore (Cambridge University Press, 1989).

Something Deeply Hidden: Quantum Worlds and the Emergence of Spacetime by Sean Carroll (Penguin, 2019).

Speakable and Unspeakable in Quantum Mechanics: Collected Papers on Quantum Philosophy by J. S. Bell (Cambridge University Press; 2nd edition, 21 June 2004).

Wholeness and the Implicate Order by David Bohm (Routledge & Kegan Paul, 1980).

Endnotes

PREFACE

1. UK National Quantum Technologies Programme (United Kingdom, UK Research and Innovation, 2024) https://uknqt.ukri.org/about-us/
2. State of Quantum 2024 (Finland, IQM Quantum Computers, January 2024) https://www.meetiqm.com/newsroom/press-releases/state-of-quantum-report-2024
3. McKinsey & Company, 'The Rise of Quantum Computing' https://www.mckinsey.com/featured-insights/the-rise-of-quantum-computing

CHAPTER 1: THE BIRTH OF THE QUANTUM CONCEPT

1. Rovelli, C., *Helgoland* (London, Allen Lane, 2020).
2. Radvanyi, P. and Villain, J., 'The discovery of radioactivity', *Comptes Rendus Physique* **18**, 544–50 (2017).

CHAPTER 3: WHAT LIES BENEATH

1. Mermin, D., 'Is the moon there when nobody looks? Reality and the quantum theory', *Physics Today* 38, No. 4, 38–47 (1985).
2. Pais, A., *Subtle is the Lord: The Science and the Life of Albert Einstein* (Oxford, Oxford University Press, 1982).
3. Feynman, R. P., Leighton, R. B. and Sands, M., *The Feynman Lectures on Physics*, Vol. 3, Chapter 1, Section 1-1, 'The Double-Slit Experiment' (Massachusetts, Addison-Wesley, 1965).

4. Rodgers, P., 'The Double-Slit Experiment', *Physics World* (Bristol, IOP Publishing, May 2015) https://physicsworld.com/a/the-double-slit-experiment/

5. Bohr, N., 'Essays 1932–1957 on Atomic Physics and Human Knowledge', reprinted as *The Philosophical Writings of Niels Bohr*, Vol. II (Woodbridge, Ox Bow Press 1987). See also 'Niels Bohr on the wave function and the classical/quantum divide', Henrik Zinkernagel, *Studies in History and Philosophy of Science Part B: Studies in History and Philosophy of Modern Physics* 53, 9–19 (February 2016).

6. Schrödinger, E., 'Die gegenwärtige Situation in der Quantenmechanik' (The present situation in quantum mechanics), *Naturwissenschaften* 23, No. 48, 807–12 (1935).

7. Heisenberg, W., *Physics and Philosophy* (New York, Harper Perennial Modern Classics, 2007).

8. Jaynes, E. T., *Complexity, Entropy, and the Physics of Information* (Massachusetts, Addison-Wesley, 1990) p. 381.

CHAPTER 4: REALITY WARS

1. Einstein, A., Podolsky, B. and Rosen, N., 'Can quantum-mechanical description of physical reality be considered complete?' *Physical Review* 47, 777–80 (1935).

2. 'Quantum entanglement and the Einstein-Podolsky-Rosen argument in quantum theory', *Stanford Encyclopedia of Philosophy* (Stanford, Stanford University, May 2001) https://plato.stanford.edu/entries/qt-epr/

3. The correspondence has been published in *The Born-Einstein letters: correspondence between Albert Einstein and Max and Hedwig Born from 1916 to 1955*, with commentaries by Max Born (London, Macmillan, 1971), p. 158. Note: this letter refers to the non-local aspects of quantum mechanics but not specifically to entanglement.

4. Von Neumann, J., 'Discussion of probability relations between separated systems', *Mathematical Proceedings of the Cambridge Philosophical Society* 31, No. 4, 555–63 (1935).

5. Full details may be found in: 'Bringing home the atomic world: Quantum mysteries for anybody', by N. D. Mermin, *American Journal of Physics* 49, 940–43 (1981). See also https://www.youtube.com/watch?v=qd-tKroLJTM

6. Bell, J. S., 'On the Einstein-Podolsky-Rosen paradox', *Physics* 1, No. 3, 195–200 (1964).

7. https://plus.maths.org/content/contextuality-most-quantum-thing
8. Peres, A., 'Incompatible results of quantum measurements', *Physics Letters A*, **151**, Nos 3–4, 107–8 (1990); Mermin, N. D., 'Simple unified form for the major no-hidden-variables theorems', *Physical Review Letters* 65, No. 27, 3373–6 (1990).
9. https://www.scientificamerican.com/article/researchers-use-quantum-telepathy-to-win-an-impossible-game/
10. Xu, J.-M. et al., 'Experimental demonstration of quantum pseudotelepathy', *Physical Review Letters* **129**, No. 5, 050402 (2022).

CHAPTER 5: QUANTUM INFORMATION MAGIC

1. https://lab.cccb.org/en/arthur-c-clarke-any-sufficiently-advanced-technology-is-indistinguishable-from-magic/
2. Ursin, R. et al., 'Quantum teleportation across the Danube', *Nature* 430 (7002), 849–50 (2004).
3. Bennett, C. H. et al., 'Teleporting an unknown quantum state via dual classical and Einstein-Podolsky-Rosen channels', *Physical Review Letters* 70, No. 13, 1895–9 (1993).
4. https://www.economist.com/technology-quarterly/2017/03/09/the-promise-of-quantum-encryption
5. Feynman, R. P., 'Simulating physics with computers', *International Journal of Theoretical Physics* 21, 467–88 (1982).
6. Deutsch, D., 'Quantum theory, the Church-Turing principle and the universal quantum computer', *Proceedings of the Royal Society A.* **400**, No. 1818, 97–117 (1985).
7. Turing, A. M., 'On computable numbers, with an application to the Entscheidungsproblem', *Proceedings of the London Mathematical Society* 2, No. 42, 230–65 (1937).
8. Aaronson, S., *Quantum Computing Since Democritus* (Cambridge: Cambridge University Press, 2013).
9. 'State of Quantum Computing: Building a Quantum Economy', *World Economic Forum Insight Report*, September 2022.
10. https://www.weforum.org/stories/2024/02/quantum-economy-blueprint-world-economic-forum/
11. Liegener, K., Oliver Morsch, O. and Pupillo, G., 'Solving quantum chemistry problems on quantum computers', *Physics Today* 77, No. 9, 34–42 (September 2024).

12. https://singularityhub.com/2024/08/15/first-post-quantum-cryptography-standards-to-guard-against-future-quantum-attacks/

13. Savage, N., 'Keeping secrets in a quantum world', *Nature* 623, S1–S3 (2023).

14. https://www.whitehouse.gov/briefing-room/statements-releases/2022/05/04/national-security-memorandum-on-promoting-united-states-leadership-in-quantum-computing-while-mitigating-risks-to-vulnerable-cryptographic-systems/

15. https://www.whitehouse.gov/wp-content/uploads/2022/11/M-23-02-M-Memo-on-Migrating-to-Post-Quantum-Cryptography.pdf

16. https://new.nsf.gov/news/nsf-national-quantum-virtual-laboratory-advances?utm_medium=email&utm_source=govdelivery

17. Turing, A., 'Computing machinery and intelligence', *Mind* LIX, No. 236, 433–60 (1950).

18. https://physicsworld.com/a/can-we-use-quantum-computers-to-make-music/

19. Abdyssagin, R.-B., *Quantum Mechanics and Avant-Garde Music: Shadows of the Void* (Springer Cham, 2024).

CHAPTER 6: SENSING THE UNSEEN

1. Bongs, K., Bennett, S. and Lohmann, A., 'Quantum sensors will start a revolution – if we deploy them right', *Nature* 617, 672–5 (2023).

2. https://www.ukri.org/news/five-hubs-launched-to-ensure-the-uk-benefits-from-quantum-future/

3. Lezeik, A., 'The most precise timekeeping device ever built', *Physics World* (27 August 2024) https://physicsworld.com/a/the-most-precise-timekeeping-device-ever-built/

4. https://www.nist.gov/pml/time-and-frequency-division/time-realization/cesium-fountain-atomic-clocks

5. https://www.priopulse.com/how-quantum-clocks-could-transform-financial-markets/

6. Sobel, D., *Longitude: The True Story of a Lone Genius Who Solved the Greatest Scientific Problem of His Time* (New York: Walker & Company, 1995).

7. https://news.mit.edu/2023/ligo-surpasses-quantum-limit-1023

8. Haocun, Y., McCuller, L., Tse, M., Kijbunchoo, N., Barsotti, L., Mavalvala, N. and other members of the LIGO Scientific Collaboration, 'Quantum correlations between light and the kilogram-mass mirrors of LIGO', *Nature* 583 (7814), 43–7 (2020).

9. Colella, R., Overhauser, A. W. and Werner, S. A., 'Observation of gravitationally induced quantum interference', *Physical Review Letters* 34 (23) 1472–4 (1975). The interpretation of this experiment turns out to be more subtle than I have described it. See *Quantum Gravity in a Laboratory?* by Huggett, N., Linnemann, N. and Schneider, M. D. (Cambridge, Cambridge University Press, 2023), Section 3.

10. Allen, M., 'Sensing gravity, the quantum way', *Physics World*, https:// physicsworld.com/a/sensing-gravity-the-quantum-way/. See also Peters, A., Chung, K. Y. and Chu, S., 'Measurement of gravitational acceleration by dropping atoms', *Physics World* (8 December 2021).

11. Lachmann, M. D., Ahlers, H., Becker, D. et al., 'Ultracold atom interferometry in space', *Nature Communications* 12, 1317, https://doi. org/10.1038/s41467-021-21622-5 (2021). See also https://phys.org/news/ 2021-04-atom-interferometry-space.html

12. https://science.nasa.gov/science-research/science-enabling-technology/ technology-highlights/quantum-technologies-take-flight/

13. Haleem, A., Javaid, M., Singh, R. P., Rab, S. and Suman, R., 'Applications of nanotechnology in medical field: a brief review', *Global Health Journal* 7 (2023), 70–77.

14. Drexler, E., *Engines of Creation: The Coming Era of Nanotechnology* (New York, Knopf Doubleday Publishing Group, 1987).

15. Kumar, R., Kumar, M. and Luthra, G., 'Fundamental approaches and applications of nanotechnology: a mini review', *Materials Today: Proceedings* 56, 3016–25 (2023) https://doi.org/10.1016/j.matpr.2022.12.172.

16. https://www.azonano.com/article.aspx?ArticleID=6213

17. https://www.birmingham.ac.uk/research/centres-institutes/human-brain-health/facilities/optically-pumped-magnetometers.

18. Aslam, N. et al., 'Quantum sensors for biomedical applications', *Nature Reviews Physics* 5, 157–69 (2023).

CHAPTER 7: QUANTUM BIOLOGY

1. Schrödinger, E., *What Is Life?* (Cambridge: Cambridge University Press, 1944).

2. Delbrück, M., 'A physicist looks at biology', *Transactions of the Connecticut Academy of Arts and Sciences* 38, 173–90 (1949).

3. Hore, P. J. and Mouritsen, H., 'The radical-pair mechanism of magnetoreception', *Annual Review of Biophysics* 45, 299–344 (2016) https:// doi.org/10.1146/annurev-biophys-032116-094545

4. Eccles, J. C., 'Do mental events cause neural events analogously to the probability fields of quantum mechanics?' *Proceedings of the Royal Society B: Biological Sciences* 240 (1299), 433–51 (1990) https://doi.org/10.1098/rspb.1990.0043

5. Liu, Z., Chen, Y.-C. and Ao, P., 'Entangled biphoton generation in myelin sheath', https://arxiv.org/pdf/2401.11682 (2024).

CHAPTER 8: THE MYTH OF THE VOID

1. Lamoreaux, S. K., 'Demonstration of the Casimir force in the 0.6 to 6 μm range', *Physical Review Letters* 78(1), 5–8 (1997) https://doi.org/10.1103/PhysRevLett.78.5

2. https://www.scientificamerican.com/article/something-from-nothing-vacuum-can-yield-flashes-of-light/

3. Ford, L. H. and Roman, T. A., '"Cosmic flashing" in four dimensions', *Physical Review D* 46(4), 1328–39 (1992) https://doi.org/10.1103/PhysRevD.46.1328.

4. Davies, P., *How to Build a Time Machine* (Penguin, 1991).

5. https://ntrs.nasa.gov/api/citations/19980201240/downloads/19980201240.pdf

6. https://www.quantamagazine.org/physicists-use-quantum-mechanics-to-pull-energy-out-of-nothing-20230222/; https://physicsworld.com/a/energy-can-be-teleported-over-long-distances-say-physicists/

7. Lambrecht, A., 'The Casimir effect: a force from nothing', *Physics World* (September 2002) https://physicsworld.com/a/the-casimir-effect-a-force-from-nothing/

8. Fong, K. Y., Li, H.-K., Zhao, R., Yang, S., Wang, Y. and Zhang, X., 'Phonon heat transfer across a vacuum', *Nature* 576(7786), 243–7 (2019) https://doi.org/10.1038/s41586-019-1800-4

9. Davies, P. C. W., 'Scalar production in Schwarzschild and Rindler metrics', *Journal of Physics A* 8(4), 609–16 (1975).

CHAPTER 9: UNIVERSE OUT OF NOTHING

1. Lemaître, G., 'The beginning of the world from the point of view of quantum theory', *Nature* **127**(3210), 706 (1931) https://www.nature.com/articles/127706b0

CHAPTER 10: WEIRDER STILL

1. Kwiat, P., Weinfurter, H., Herzog, T., Zeilinger, A. and Kasevich, M. A., 'Interaction-free measurement', *Physical Review Letters* 74(24), 4763–6 (1995) https://doi.org/10.1103/PhysRevLett.74.4763.

2. Vanner, M. R. et al., 'Something from nothing: enhanced laser cooling of a mechanical resonator via zero-photon detection', https://arxiv.org/pdf/2408.01734 (2024).

3. Elitzur, A. C. and Vaidman, L., 'Quantum mechanical interaction-free measurements', *Foundations of Physics* 23(7), 987–97 (1993) https://doi.org/10.1007/BF00736012

4. Wheeler, J. A., 'The "past" and the "delayed choice" double-slit experiment'. This paper first appeared in 1978 and has been reprinted in several locations, e.g. Lisa M. Dolling, Arthur F. Gianelli, Glenn N. Statilem, *Readings in the Development of Physical Theory*, pp. 486ff.

5. Jacques, V., Wu, E., Grosshans, F., Treussart, F., Grangier, P., Aspect, A. and Roch, J.-F., 'Experimental realization of Wheeler's delayed-choice gedanken experiment', *Science* **315**(5814), 966–8 (2007) https://doi.org/10.1126/science.1136303

6. Scully, M. O. and Drühl, K., 'Quantum eraser: A proposed photon correlation experiment concerning observation and "delayed choice" in quantum mechanics', *Physical Review A*, **25**(4), 2208–13 (1982) https://doi.org/10.1103/PhysRevA.25.2208

7. Kim, Y.-H., Yu, R., Kulik, S. P., Shih, Y. H. and Scully, M. O., 'A delayed "choice" quantum eraser', *Physical Review Letters* 84(1), 1–5 (2000) https://doi.org/10.1103/PhysRevLett.84.1

8. See ref. 7. See also https://www.scientificamerican.com/article/quantum-eraser-delayed-choice-experiments/

9. Lloyd, S., 'Quantum procrastination', *Science* 338(6107), 621–2 (2012) https://doi.org/10.1126/science.1231471

10. Hertog, T., *On the Origin of Time: Stephen Hawking's Final Theory* (Bantam, 2023).

11. Davies, P., *The Goldilocks Enigma: Why is the universe just right for life?* (Allen Lane, 2006).

12. Denkmayr, T., Geppert, H., Sponar, S. et al., 'Observation of a quantum Cheshire Cat in a matter-wave interferometer experiment', *Nature Communications* **5**, 4492 (2014) https://doi.org/10.1038/ncomms5492.

13. Haldane, J. B. S., *Possible Worlds and Other Essays* (Chatto & Windus, 1927).

CHAPTER 11: MAKING SENSE OF IT ALL

1. Adler, S. L. and Bassi, A., 'Is quantum theory exact?' *Science* 325(5938), 275–6 (2009) https://doi.org/10.1126/science.1177116

2. https://phys.org/news/2022-06-collapsing-theory-quantum-consciousness.html

3. Wheeler, Chapter 10, reference 4.

4. Hameroff, S. and Penrose, R., 'Consciousness in the universe: a review of the "Orch OR" theory', *Physics of Life Reviews* 11(1), 39–78 (2014) https://doi.org/10.1016/j.plrev.2013.08.002.

5. Leggett, A. J., 'Reflections on the quantum measurement paradox', in Hiley, B. J. and Peat, F. D. (eds), *Quantum Implications: Essays in Honour of David Bohm* (London, Routledge, 1991), 85–104.

6. Popescu, S. and Rohrlich, D., 'Nonlocality as an axiom', *Foundations of Physics* 24(3), 379–85 (1994) https://doi.org/10.1007/BF02058098; Popescu, S., 'Nonlocality beyond quantum mechanics', *Nature Physics* 10, 264–70 (2014) https://doi.org/10.1038/nphys2916

7. Schlosshauer, M., *Decoherence and the Quantum-to-Classical Transition* (Springer, 2007).

8. Hossenfelder, S., 'Superdeterminism: a guide for the perplexed', https://arxiv.org/pdf/2010.01324.pdf

9. Palmer, T., 'Superdeterminism without conspiracy', https://doi.org/10.3390/universe10010047

10. Fuchs, C. A., Mermin, N. D. and Schack, R., 'An introduction to QBism with an application to the locality of quantum mechanics', *American Journal of Physics* 82(8), 749–54 (2014) https://doi.org/10.1119/1.4874855

11. Di Biagio, A. and Rovelli, C., 'Stable facts, relative facts', *Foundations of Physics* 51(1), 30 (2021) https://doi.org/10.1007/s10701-021-00429-w

12. Schmidt, D., Spekkens, R. W. and Shelby, J. H., 'Unscrambling the omelette of causation and inference: the framework of causal-inferential theories', https://arxiv.org/pdf/2009.03297

13. Wheeler, Chapter 10, reference 4, p. 357.

14. Brown, J. R. and Davies, P. C. W., *The Ghost in the Atom* (Cambridge, Cambridge University Press, 1986).

15. Bohr, N., *Atomic Physics and Human Knowledge* (New York, Wiley, 1958).

Acknowledgements

Many people have helped me over the years with the subject matter of this book. Special thanks must go to Scott Aaronson, Yakir Aharonov, Clarice Aiello, Andrew Briggs, David Deutsch, Darren Dougan, Avshalom Elitzur, George Ellis, Baruch Garcia, Andrew Jordan, Seth Lloyd, Chiara Marletto, Tim Palmer, Roger Penrose, Sandu Popescu, Gerard Milburn, Kanu Sinha, Louie Slocombe, Phil Tee, Jeff Tollaksen and Lev Vaidman. Pauline Davies provided very detailed critical feedback on earlier drafts, and the book is greatly improved by her input. I should also like to thank my editor, Chloe Currens, for her commitment to this project and her perceptive critiques and suggestions.

February 2025

Index